REMOTE BRAIN TARGETING

EVOLUTION OF MIND CONTROL IN USA

A COMPILATION OF HISTORICAL INFORMATION
DERIVED FROM VARIOUS SOURCES

Renee Pittman

Remote Brain Targeting

Copyright © 2012 by Renee Pittman. All rights reserved.

Mother's Love Publishing and Enterprises

ISBN-13: 978-1-7374060-9-9

This book is not designed to provide authoritative information in regard to the subject matter covered. This information is given with the understanding that the author is not engaged in rendering legal or professional advice. Since the details of any situation are fact dependent, you should additionally seek the services of a competent professional.

DEDICATION

To AWARENESS and to many who are
first discredited then suffer unjustly, COVERTLY

Table of Contents

Chapter One: Mind Control Technology 1

Chapter Two: Mind Control Techniques 130

Chapter Three: Microwave Hearing Schizie 157

Chapter Four: Behavior Modification 212

Chapter Five: Psychotronics 224

Chapter Six: Research 243

Chapter Seven: History..255

Chapter Eight: Testing Programs 266

Chapter Nine: Timeline 277

Chapter Ten: Mass Population Control 348

Chapter Eleven: Conclusion....................................402

References 429

CHAPTER ONE

"The object of terrorism is terrorism. The object of oppression is oppression. The object of torture is torture. The object of murder is murder. The object of power is power. Now do you begin to understand me?"

– George Orwell (1984)

The object of mind control is mind control.

– Pittman (2013)

On June 13, 1975 in a major speech, Leonid Brezhnev, General Secretary of the Central Committee of the Communist Party of the Soviet Union, in office 1964 to 1982, urged the United States to agree on a ban of research and development of new kinds of weapons "more terrible than anything the world has ever known. On July 2, 1975, Brezhnev again, repeated this proposed ban on the development of frightful new weapons, to a group of United States Senators. In August of 1975, Ponomarev called for a ban on frightful new weapons of mass destruction. On September 23, 1975, Gromyko presented a draft treaty agreement to the 30th Session of the UN General Assembly, for banning development of frightful new weapons.

The first phase of government mind control development grew out of the old occult techniques which required the victim to be exposed to massive psychological and physical trauma, usually beginning in infancy, in order to cause the psyche to shatter into a thousand alter personalities which can then be separately programmed to perform any function (or job) that the programmer wishes to "install". Each alter personality created is separate and distinct from the front personality. The 'front personality' is unaware of the existence or activities of the alter personalities. Alter personalities can be brought to the surface by programmers or handlers using special codes, usually stored in a laptop computer. The victim of mind control can also be affected by specific sounds, words, or actions known as triggers. Dr. Stefan Possony, in the beginning of his writing "In Scientific Advances Hold Dramatic Prospects for Psy-Strategy," by Dr. Stefan Possony, 1983 July, Defense and Foreign Affairs, Page 34, Dr. Possony states: "The history of psy-ops technology is about 200 years old and it will continue to progress."

Dr. Possony was accurate. In the 1700s some of the first testing began as man began to make the connection between electricity and the human brain and mind. In the history of "Electrical Excitation," Drs. Chaffee and Light in, "Electrical Stimulation of the Nervous System, 1934" wrote of the first awareness of electromagnetism and the human body saying, "The history of neurophysiology has been decided in large part by the development of electric-recording instruments on the one hand, and by the increasingly effective use of electric currents for stimulating on the other."

The early awareness of this connection would set the stage for future global goals for mass population mind control in a monstrous program unsurpassed.

IF A LOT OF PEOPLE ARE SAYING THE SAME THING – LISTEN!

Known as E. Howard Hunt, Everette Howard Hunt, Jr. was an American intelligence officer, insider, and writer. He served as a CIA officer from 1949 to 1970. Along with G. Gordon Liddy and others, Hunt was one of the Nixon White House "plumbers."

The plumbers were a secret team of operatives charged with fixing "leaks" (real or perceived causes of confidential Administration information being leaked to outside parties). Hunt and Liddy engineered the Watergate burglaries and other undercover operations for the Nixon Administration. In the ensuing Watergate scandal; Hunt was convicted of burglary, conspiracy, and wiretapping, eventually serving 33 months in prison.

Because of a secret contract between the CIA and highly trained Scientologists at the inception of the CIA-initiated remote viewing program, events related to Scientology and L. Ron Hubbard that led up to the unlikely marriage of CIA and Scientology are included.

TIMELINE ONE
Scientology and the Remote Viewing Connection

NOTE: This timeline, and anything connected to Scientology over the years, has raised a lot of controversy. However, many report having their lives transformed in powerful ways by this religion. Yet others, have reported disillusion which can be said for any religion as well.

The timeline, documented here, has no personal interest, nor personal feelings related either, positively or negatively, to Scientologist beliefs. This timeline simply hopes to connect history and the role Scientology has played globally.

In summary, it appears that a take down effort materialized when L. Ron Hubbard refused involvement with intelligence agency objectives and his later exposing their agendas.

1940s

October 1943

E. Howard Hunt is confirmed for service in Office of Strategic Services

CIA's E. Howard Hunt

OSS), forerunner to CIA. Hunt goes to Catalina Island for training. Among the people Hunt trains with is Lucien Conein. [1]

1946

Psychiatrist Lewis J. Fielding joins the Veteran's Administration (VA) in Los Angeles as staff psychiatrist and instructor in clinical psychiatry. When Fielding joins the VA, and intermittently over the next several years, Dianetics and Scientology Founder L. Ron Hubbard is living in and around L.A., doing Dianetic research. He has opened an office in Hollywood, California, and is delivering Dianetic processing to people. When there, he is associated with the Los Angeles Veteran's Administration.

[NOTE: Fielding later will be integral to incidents involving CIA's E. Howard Hunt, Daniel Ellsberg, Lucien Conein, the Pentagon Papers, and the Watergate scandal—at the very time that CIA is secretly setting up its remote viewing program using highly trained Scientologists. See timeline years 1971 and 1972.] [2][3][4]

Thursday, 18 September 1947

The National Security Act of 1947 establishes the Central Intelligence Agency (CIA). Robert Komer is in the CIA at its inception.

[NOTE: Komer will become a close associate of Daniel Ellsberg at Rand and will be with Ellsberg in Vietnam.] [5][6]

Wednesday, 15 October 1947

A letter purportedly from L. Ron Hubbard, dated 15 October 1947, ends up in his Veteran's Administration files in Los Angeles, where Lewis J. Fielding is staff psychiatrist. The subject of the letter is Hubbard pleading with the VA for psychiatric help. [

NOTE: The letter never surfaces until decades later, and then only as a copy several generations removed, making its authenticity impossible to prove or disprove.] [7]

June 1948

With CIA liaison for administration and supply, a separate clandestine organization called Office of Policy Coordination (OPC) is created. On CIA orders, E. Howard Hunt reports to Washington, D.C. to begin service in OPC. [1]

July 1949

CIA's head of Scientific Intelligence goes to Western Europe to learn Soviet techniques in mind control and interrogation, including use of LSD. [8]

1950s

Thursday, 20 April 1950

CIA's mind control program Project BLUEBIRD is authorized. CIA-contracted psychiatrists begin secret experiments with icepick lobotomies, electroshock, hypnosis, pain, and drugs, including cocaine, heroin, and LSD. In coordination with the Veteran's Administration, U.S. military veterans are used as unwitting subjects for many of the experiments. [9][8]

Tuesday, 9 May 1950

The book Dianetics, the Modern Science of Mental Health by L. Ron Hubbard is released. It decries hypnosis, and describes techniques for safely accessing in the mind the contents of incidents involving unconsciousness, hypnosis, drugs, and pain. It becomes a bestseller. [10] [11]

Saturday, 27 May 1950

With Dianetics a bestseller, L. Ron Hubbard (right) is lecturing around the country when U.S. Naval Intelligence attempts to force him into service for the U.S. government. He refuses.

The Office of Naval Intelligence in Washington, D.C. sends an officer to put L. Ron Hubbard into civilian service in the government to continue his researches on the mind. Hubbard says no. The officer says that if he refuses, Hubbard will be ordered back to active duty, since his Naval commission has not been terminated. Hubbard quickly takes advantage of a letter of permission he has from the Secretary of the

Navy to resign his commission, thereby putting Dianetics and Scientology out of the reach and control of the U.S. government. [12] [13]

September 1950

As CIA's Project BLUEBIRD expands, the CIA-contracted psychiatrists' experimental purposes and activities include inducing amnesia, inserting hypnotic access codes in subjects' minds, controlling behavior from remote transmitters with brain electrodes, administering LSD to children, and using electroshock to erase memories. [9]

1951

Office of Policy Coordination (OPC), where E. Howard Hunt is working, is merged with CIA. Over the coming years, during his CIA career, Hunt has other occasions to work with Lucien Conein, who is on contract to CIA. [1]

January 1951

L. Ron Hubbard introduces the "Theta-MEST Theory" stating that thought (Theta) is separate from the physical universe (Matter, Energy, Space and Time—MEST): that Theta can operate in and with MEST, that Theta can consider itself integrated with MEST, and that Theta can consider itself to be MEST, but that creative thought and perception reside in Theta, not MEST. [14]

Monday, 25 June 1951

L. Ron Hubbard exposes "a carefully guarded secret of certain military and intelligence organizations."

CIA's Sidney Gottlieb

In a new book, Science of Survival, Hubbard says: "It required Dianetic processing to uncover pain-drug-hypnosis. Otherwise, pain-drug-hypnosis was out of sight, unsuspected, and unknown." Hubbard denounces its use as a "vicious war weapon" that may be "of considerably more use in conquering a society than the atom bomb." [NOTE: It's not until decades later that CIA's pain-drug-hypnosis experimentation during this period begins to be investigated and reported by Congress. By that time, CIA's Richard Helms, Sidney Gottlieb, and others will have destroyed many of CIA's records of such activities. See January 1973.] [15]

Monday, 20 August 1951

CIA's Project BLUEBIRD evolves into Project ARTICHOKE, with goals such as "get control of an individual to the point where he will do our bidding against his will and even against fundamental laws of nature."[16]

Monday, 7 January 1952

A secret internal CIA document discusses a multi-level program to research and develop the use of extrasensory perception for "practical problems of intelligence."[17]

July 1952

L. Ron Hubbard releases a book, History of Man (published also as What to Audit), that describes some of the native capabilities of thought (Theta) in the individual as including communication by telepathy and the moving of material objects by "throwing an energy flow at them." Hubbard describes Scientology processes to rehabilitate these potentials. [18]

Wednesday, 6 August 1952

Alexander Puharich delivers a lecture called "On the Possible Usefulness of Extrasensory Perception in Psychological Warfare" to a Pentagon conference. [19]

"On the Possible Usefulness of Extrasensory Perception in Psychological Warfare" becomes a decades-long intelligence agenda for the Cold War.

December 1952

George Hunter White, on loan to CIA from the Federal Narcotics Bureau, begins administering LSD to unwitting U.S. citizens at a CIA "safe house" in Greenwich Village. [20]

L. Ron Hubbard delivers a series of over 50 lectures in Philadelphia on processes for attaining a state he calls "Operating Thetan" (OT), described as a being stably exterior from the body and able to perceive, communicate, and operate in the physical universe without reliance on the sense channels or mechanics of a body. [21]

1953

James McCord, later to be involved in the Watergate break-in, joins CIA. [22]

Friday, 10 April 1953

CIA director Allen Dulles gives a speech before the National Alumni Conference at Princeton University, lecturing on "how sinister the battle for men's minds" has become in Soviet hands. [23]

Monday, 13 April 1953

CIA's Richard Helms

CIA Director Allen Dulles authorizes a new expanded mind-control program, MK-ULTRA, brainchild of Richard Helms, a high-ranking member of CIA's Clandestine Services. E. Howard Hunt is working at CIA headquarters at the time as a "chief of covert operations" under Clandestine Services. [23] [24] [1]

Thursday, 19 November 1953

On a 3-day holiday for CIA officials at Deer Creek Lodge in the mountains of Maryland, Sidney Gottlieb—head of CIA's MK-ULTRA—secretly slips LSD into the after-dinner drinks. An Army scientist and germ warfare specialist named Frank Olson, who is working on a MK-ULTRA project, experiences a "bad trip," becoming very disoriented.

[NOTE: Olson soon commits suicide.] [20]

Circa July 1955

George Hunter White moves his CIA "safe house" operation, equipped with one-way mirrors and surveillance gadgets, to San Francisco, under the aegis of MK-ULTRA and Sidney Gottlieb. The code name is Operation Midnight Climax. He hires prostitute addicts who lure men from bars to the safe houses after their drinks have been spiked with LSD. White films the events. The purpose of these "national security brothels" is to enable CIA to experiment with the act of lovemaking for extracting information from men. [25]

Factions of the U.S. government are making efforts to "seize Scientology in the United States."[26]

Monday, 15 August 1955

Staff of CIA Director Allen Dulles complete "A Report on Communist Brainwashing."[27]

Wednesday, 2 January 1957

The Church of Scientology of California, the senior church, is granted tax exemption. [28]

July 1957

The CIA has file No. 156409 on L. Ron Hubbard and his organizations. [29]

January 1958

L. Ron Hubbard introduces the "American Blue E-meter," a transistorized improvement over earlier prototypes, to be used as an aid for Scientology practitioners. Newer Scientology technology begin to require the use of the meter as a guide to the use of processes toward the attainment of Operating Thetan (OT). [11]

June 1958

Daniel Ellsberg arrives at Rand to spend the summer as a consultant. [30]

Circa July 1958

U.S. corporations, including Westinghouse, General Electric, and Bell Telephone have begun telepathy research. [17]

Saturday, 8 November 1958

The Herald Tribune in New York reports that Westinghouse Electric Corporation has begun to study ESP using specially designed apparatus. [17]

L. Ron Hubbard discovers measurable sentience in plants, first using an E-meter with geraniums in his greenhouse at St. Hill, England, later with tomatoes

Saturday, 25 July 1959

Westinghouse Corporation's Friendship Laboratory undertakes an experiment in ESP with the U.S.S. Nautilus, linking one person on land (the sender) with another person in the submarine (the receiver), while the vessel is submerged. Representatives of the U.S. Navy and Air Force are present during the experiments, which run for sixteen days under Air Force Colonel William H. Bowers. The experiments result in a 70% success rate. [17]

Friday, 18 December 1959

Garden News publishes a story, "Plants Do Worry and Feel Pain," describing experiments done by L. Ron Hubbard where he has connected plants to a Scientology E-meter and measured their reaction to threat and damage. [31]

1960s

Monday, 17 April 1961

Martin Ebon, administrative assistant of the Parapsychology Association in New York and on staff with the U.S. Information Agency, is in Washington, D.C. giving a briefing on telepathy to "a top intelligence agency." He is a specialist in tracking and examining the nature and direction of Russian and Soviet security services. [17]

Tuesday, 25 July 1961

L. Ron Hubbard goes from St. Hill in England to Cape Finisterre, Spain for about 10 days and purchases a ship that is 106 feet long with an 18-foot beam and sleeps about 22 people. He says that purchase of the ship is "part of an operation" he is conducting from St. Hill.

[NOTE: This event is the earliest indication of Hubbard's creation, in Spain, of the confidential "Sea Project," later to be called the "Sea Org," which he establishes to protect and deliver the confidential upper levels of Scientology. This is the first ship purchased. The Sea Project will not be made known publicly for several years. Later, Hubbard writes that the "oldest yacht in the Sea Org" is the "Enchanter."] [32] [33]

Tuesday, 1 August 1961

The Defense Intelligence Agency (DIA) is created in the Department of Defense (DOD) by DOD Directive 5105.21, August 1, 1961.

[NOTE: After the 1971-1972 creation of the CIA-initiated remote viewing program, its administration later gets transferred, at least in part, to DIA as one action of the remote viewing shell-game of secrecy that goes on for decades.]

The Soviet Union claims to have confirmed "communication between two people separated by long distances...without conventional communication facilities."

July 1962

A book published in the Soviet Union, "Biological Radio Communication," claims "experimental confirmation of the fact that communication between two people separated by long distances can be carried out through water, over air and across metal barrier by means of cerebral radiation in the course of thinking, and without conventional communication facilities."[17]

1963

CIA organizes the Robert R. Mullen public relations firm as a CIA front company for use "mainly in providing CIA cover overseas." Mullen has several overseas branches for its CIA front operations and an office in Washington, D.C. One of its branches in Europe is staffed, run, and paid for by CIA.

[NOTE: E. Howard Hunt will "retire" from CIA and go to work for Mullen on 1 May 1970 (see). At the time of his purported retirement, Hunt will be CIA's "Chief EUR/CA"] [34]

Feds Seize Scientology / E-meters

CIA's Richard Helms attempts to defend CIA drug operations like Midnight Climax by telling CIA Inspector General John Earman that such testing has been necessary "to keep up with the Soviets."[25]

Friday, 4 January 1963

Deputy Marshals and Food and Drug Administration Agents raid the Scientology headquarters in Washington, D.C. and seize 100 E-meters along with Scientology publications on the grounds that the E-meters are "misbranded."[35]

1964

Daniel Ellsberg is granted a unique security clearance giving him "unprecedented access to data and studies in all agencies," as well as several "special clearances" including "very high-level access to all our secrets in the State and Defense Departments and the CIA."[2]

Circa November 1964

CIA Director Helms testifies. He later will be convicted of perjury for lying to Congress.

Richard Helms, Director of CIA and father of MK-ULTRA, contradicts himself on the need "to keep up with the Soviets," telling the Warren Commission that "Soviet research has consistently lagged five years behind Western research."[25]

Early January 1965

L. Ron Hubbard and Mary Sue Hubbard leave St. Hill in England and travel to Spain, going first at the Canary Islands—a province of Spain—off the coast of Morocco. His research into upper levels of Scientology is now confidential, and while in Spain and the Canary Islands, he lays the groundwork for establishing confidential bases in Spain to deliver the upper levels. They return to St. Hill mid-February 1965. [3] [36]

Monday, 17 May 1965

Psychiatrist Jose Delgado gives a mind control demonstration in Spain with a bull: he presses a button on a radio transmitter and the bull brakes to a halt. He presses another button and the bull turns to the right and trots away, obeying commands being delivered through radio signals to brain regions in which fine wires had been planted the day before. [25]

Monday, 14 June 1965

L. Ron Hubbard issues "Politics, Freedom From" in Executive Directive form, declaring Scientology to be "nonpolitical and non-ideological," and declaring it "free of any political connection or allegiance of any kind whatever." He says the reason for the declaration is the continuing efforts of the U.S. government "to seize Scientology in the United States." In closing he says: "Scientologists may be members of any political group on this planet without restraint only so long as these individuals or that group do not attempt to seize Scientology for their own warlike ends and so make it unworkable or distasteful by invidious connection."[26]

Tuesday, 29 June 1965

L. Ron Hubbard tells a group of Scientology students in a lecture at St. Hill in England that he has recently returned from Washington D.C. where IRS and factions of the U.S. government have been "trying to seize Scientology in the United States," and that he has told them an emphatic "no."[37]

July 1965

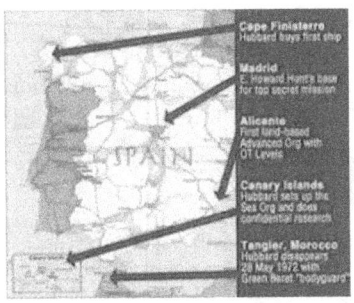

Hunt and Hubbard in Spain

E. Howard Hunt converts from "CIA employee" to "CIA contract status" and is sent by Richard Helms on a secret mission to Madrid, Spain. The Canary Islands, where L. Ron Hubbard has recently begun to establish a base for upper level Scientology research, is a province of Spain at the time. [38] [1]

In a meeting with CIA's William Colby it is arranged that Daniel Ellsberg will be traveling to Vietnam with E. Howard Hunt's long-time associate, CIA's Lucien Conein, on a team headed up by 20-year CIA veteran Edward Lansdale [often misspelled as Edward Landsdale]. [30]

Friday, 3 September 1965

L. Ron Hubbard inaugurates the confidential Clearing Course, available only at St. Hill in East Grinstead, Sussex, England. [3]

Monday, 11 October 1965

Moscow's A.S. Popov Scientific-Technical Society for Radio Engineering, Electronics and Communication establishes a Laboratory for Bio-Information to conduct laboratory-controlled telepathic experiments.

[NOTE: Also referred to as Soviet Laboratory for Bio-Electronics and the Laboratory for Bio-Communications.] [17]

Sunday, 17 October 1965

Daniel Ellsberg has arrived in Vietnam with CIA's Edward Lansdale [often misspelled as Edward Landsdale] and Lucien Conein, and is put in touch with John Paul Vann. Through Vann, Ellsberg meets reporter Neil Sheehan. For about six weeks Vann drives Ellsberg around to every province capital in "III Corps"—the eleven provinces that include Saigon.

[NOTE: Neil Sheehan will later publish secret documents given to him by Ellsberg: the "Pentagon Papers.] [30]

Tuesday, 14 December 1965

Stephen I. Abrams, Director of the Parapsychological Laboratory, Oxford University in England, working under the auspices of CIA's MK-ULTRA, prepares a review article entitled "Extrasensory Perception." It says ESP has been demonstrated, but is not understood or controllable. [19]

L. Ron Hubbard forbids access to confidential Scientology upper levels for anyone connected to "police spy organizations and government spy organizations" including CIA, IRS, FBI, an NSA.

Tuesday, 28 December 1965

L. Ron Hubbard issues a Scientology policy letter that forbids anyone connected to a "Suppressive Group" from being allowed onto the confidential Scientology upper levels unless and until the group is permanently disbanded. "Suppressive Groups" are defined as those that "seek to destroy Scientology" or specialize in "injuring or killing people or damaging their cases," or that "advocate the suppression of Mankind." They include "police spy organizations and government spy organizations" such as the CIA, IRS, FBI, National Security Agency (NSA), Department of Justice (DOJ), "or any other federal agency in any country."[39] [40]

Wednesday, 2 February 1966

CIA's Cleve Backster plagiarizes plant response testing, giving CIA their own data set for experiments done years earlier by Hubbard that he would never share with an intelligence agency

CIA contractor Cleve Backster connects plants to polygraphs and gets reactions on the machines when the plants are threatened or harmed.

[NOTE: These are almost identical to plant experiments done by L. Ron Hubbard over six years earlier using the Scientology E-meter. See 18 December 1959.] [41]

March 1966

Daniel Ellsberg is working in Vietnam in conjunction with CIA's Intelligence Coordination and Exploitation (ICEX) operations—forerunner of the Phoenix Project—on "hamlet pacification," a euphemism for, among other things, kidnapping, brutal interrogations, and assassinations. [2] [30]

Thursday, 7 April 1966

Daniel Ellsberg is in a meeting in Vietnam with his "old friend from Rand," CIA veteran and National Security Council (NSC) member Robert Komer. At the time, Ellsberg is involved with Lucien Conein and John Paul Vann. One of the Green Berets in the various CIA "pacification programs" is Paul Preston.

[NOTE: Preston later will enroll in Scientology, and will be described by some sources as having become the "bodyguard" of L. Ron Hubbard when Hubbard disappears. See 28 May 1972.] [2] [30]

Circa July 1966

"Special Department No. 8" is established at the Institute of Automation and Electrometry in Academgorodok, ("Science City"), near Novosibirsk, Siberia. The building that houses the department can only be entered if one knows the code, changed each week, that opens the main door's lock. The "No. 8" operation is devoted to experiments in information transmission by "bioenergetics" means. About 60 researchers have been brought to the facility from other parts of the USSR. [17]

Saturday, 9 July 1966

The Moscow daily Komsomolskaya Pravda reports on long-distance telepathy experiments conducted by the Moscow Laboratory of Bio-

Information, using Yuri Kamensky and Karl Nikolayev. The experiments are reported to have "demonstrated the reality of the phenomenon and produced valuable data, both positive and negative, which pointed up the need for continued research."[17]

The Soviet Union reports more successful telepathy experiments, escalating the Cold War race for supremacy in psychological warfare.

Friday, 29 July 1966

IRS sends a letter to the senior Scientology organization, Church of Scientology of California, recommending revocation of tax exemption. [28]

Sunday, 14 August 1966

The confidential materials for the first Operating Thetan level, OT I, are released by L. Ron Hubbard to qualified Scientologists on 14 August 1966. The level is called OT I. [11]

Tuesday, 16 August 1966

L. Ron Hubbard issues a Scientology policy letter called "Clearing Course Security" with instructions on how to handle reports of anyone being a "potential security risk" with confidential upper level materials. [42]

Thursday, 1 September 1966

L. Ron Hubbard resigns from all directorships and running of Scientology organizations. At about the same time he releases to qualified Scientologists the confidential materials of "Operating Thetan Level II" (OT II). [11]

Wednesday, 21 September 1966

CIA's E. Howard Hunt returns to the Washington, D.C. area from a highly secret assignment he has been on in Spain for a little over a year. Hunt supposedly has been on a "contract" basis with CIA rather than an employee of CIA since leaving for Spain, but a CIA document of 21 September, sent to CIA's Central Cover Staff through the Office of Security refers to Hunt as "this employee."

[NOTE: Also see 1970, where Hunt purportedly "retires" from CIA as an employee.] [1] [38]

Wednesday, 5 October 1966

Just after E. Howard Hunt arrives in D.C., Daniel Ellsberg leaves Vietnam and flies to Washington, D.C., then turns around almost immediately and makes a one week trip back to Vietnam, flying non-stop on the plane of Secretary of Defense Robert McNamara. Ellsberg is in contact again with CIA's Robert Komer. [30]

Wednesday, 12 October 1966

Daniel Ellsberg flies back from Vietnam to Washington, D.C. non-stop on McNamara's plane. With them is CIA's Robert Komer. [30]

Thursday, 10 November 1966

L. Ron Hubbard has created the "Sea Project" and confidential programs for Scientology Clears and OTs to carry out. [43]

Tuesday, 29 November 1966

Harold "Hal" Puthoff from NSA and Ingo Swann from the UN enroll in Scientology, supposedly unknown to each other. Within five years they will be at the highest levels of Scientology and under secret contract with CIA to develop remote viewing for military intelligence.

L. Ron Hubbard gives a lecture to the Saint Hill Special Briefing Course: "OT and Clear defined," date code 6611C29, lecture code SHspec-82.

1967

Ingo Swann begins to take Scientology services. At about the same time, Swann tenders his two-year notice for resignation from his permanent contract with the United Nations Secretariat in New York. [44]

NSA's Hal Puthoff enrolls in Scientology services.

[NOTE: Puthoff will somehow get past or around the Hubbard injunction against members of a "Suppressive Group" being allowed access to the upper levels of Scientology, and by 1971 Puthoff will have attained the highest level, OT VII. See January 1971.]

July 1967

IRS revokes the tax exemption of the senior Scientology organization, Church of Scientology of California (CSC), which it has had since 2 January 1957 (see). [28]

Wednesday, 10 January 1968

L. Ron Hubbard reissues "Politics, Freedom From" [see 14 June 1965], this time as a broad public issue Hubbard Communication Office Policy Letter, declaring Scientology to be "nonpolitical and nonideological," and declaring it "free of any political connection or allegiance of any kind whatever."

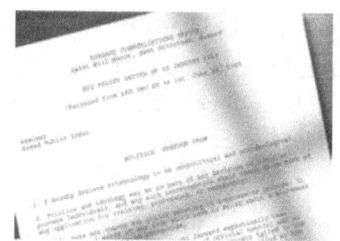

L. Ron Hubbard issues "Politics, Freedom From" as policy

He says the reason for the declaration is the continuing efforts of the U.S. government "to seize Scientology in the United States."[26]

February 1968

"American intelligence analysts" begin "noticing a Soviet secret police (KGB) trend...indicating serious interest in what is called 'parapsychology' in the West."[17]

The first Scientology "Advanced Org" is started on the Scientology Flagship (then called the Royal Scotman, later the Apollo) for the delivery of the confidential upper levels. The location is highly confidential. [45]

March 1968

Daniel Ellsberg is working at Rand Corporation in Santa Monica, California. He begins having regular meetings with Beverly Hills psychiatrist Lewis J. Fielding—former staff psychiatrist of the Veteran's Administration when L. Ron Hubbard was connected with the VA there. Ellsberg purportedly is a Fielding patient. [2]

Tuesday, 4 March 1969

On a trip to the Rand office in D.C. Daniel Ellsberg is given a copy of the top-secret study that will become known as "the Pentagon Papers" for him to carry with him to the Rand office in Santa Monica, California where he works. [30]

April 1969

Ingo Swann employment on a "permanent contract" with the United Nations Secretariat in New York comes to an end. [44]

Tuesday, 30 September 1969

G. Gordon Liddy gets "special clearances" from CIA

Daniel Ellsberg "decides to release the Pentagon Papers," purportedly because he reads in the newspaper that all charges have been dropped against several Green Berets who had been charged with a murder in Vietnam.

[NOTE: The kind of ops the Green Berets had been charged with were what Ellsberg himself had been immersed in while in Vietnam, including time spent with his "good friend" Robert Komer—who headed CIA's Phoenix Project of kidnappings and assassinations.] [30]

Tuesday, 16 December 1969

A Scientology Guardian Order says that double agents are being infiltrated into Scientology staffs and urges the use of any means to detect such infiltration. [46]

December 1969

G. Gordon Liddy is granted "special clearances" by CIA. [34]

1970

IRS's Meade Emory

January 1970

Meade Emory is Legislation Attorney for the Joint Committee on Taxation of the U.S. Congress.

[NOTE: Emory has ties to the law firm Gall, Lane, Powell and Kilcullen in D.C. Emory will later be Assistant to Commissioner of IRS, then secretly will be the architect of several wills attributed to L. Ron Hubbard, and of the restructuring of all Scientology corporations, turning control of all Scientology materials over to non-Scientology tax attorneys appointed for life. See Meade Emory, Founder.] [47]

Sunday, 22 February 1970

Yvonne Gillham, one of the earliest Sea Org members, has been Commanding Officer of the Scientology Advanced Organization in Los Angeles (AOLA). L. Ron Hubbard gives her a mission to set up a Scientology organization called Celebrity Centre in Los Angeles, with the purpose of "revitalization of the arts." She rents an old redwood and brick warehouse near downtown Los Angeles and begins delivering basic Scientology services to people in the arts, sports, entertainment, and government. She knows Ingo Swann, and he is instrumental in helping her get Celebrity Centre started.

Yvonne Gillham (later Jentzsch) opens Celebrity Centre in Los Angeles

The center also has an art gallery, a performance theatre, and classes in everything from ballet to fencing to fine arts and crafts like pottery and leatherwork.

Friday, 10 April 1970

Richard Helms has rubber-stamped E. Howard Hunt's "early retirement" and has written a letter to Robert R. Mullen on behalf of Hunt, urging Mullen to hire him. Mullen is head of a CIA-created public relations firm in D.C. and has "cooperated" with CIA in the past." One of the Mullen offices, in Stockholm, Sweden, is "staffed, run, and paid for by CIA." Also, at the Mullen firm is Douglas Caddy. [1] [38]

Monday, 13 April 1970

Daniel Ellsberg quits Rand in California, flies to Boston and signs a contract at MIT. He remains, though, a "consultant" for Rand. [30]

Friday, 1 May 1970

E. Howard Hunt ostensibly "retires" from CIA. He goes to work for Mullen in D.C.

[NOTE: By this time, up to eight people at the Mullen company have been "cleared and made witting of Agency ties, mainly in providing CIA cover overseas." Some time shortly after arriving, Hunt is told by Robert Mullen that Mullen is planning to retire before long, and that Douglas Caddy has been selected to run the CIA front company along with Hunt and an unnamed other person after Mullen's retirement.] [1] [48]

Four weeks after Hunt's purported retirement from CIA and employment at Mullen, a CIA Covert Security Approval is requested for Hunt under Project QK/ENCHANT.

Tuesday, 5 May 1970

Daniel Ellsberg flies to Washington, D.C. and is there for three days, flies to St. Louis for a day, then flies back to D.C. [30]

Wednesday, 13 May 1970

Daniel Ellsberg testifies before the Senate Foreign Relations Committee hearings, Senator Fulbright chairing. [30]

Thursday, 28 May 1970

A CIA Covert Security Approval is requested under Project QK/ENCHANT for E. Howard Hunt. [49]

August 1970

CIA's James McCord

Just four months after E. Howard Hunt, James McCord "retires" from CIA. [50]

September 1970

Daniel Ellsberg stops seeing Beverly Hills psychiatrist Lewis Fielding.

[NOTE: Ellsberg had moved to Cambridge, Massachusetts months before. See 13 April 1970.] [2]

Sunday, 20 September 1970

L. Ron Hubbard releases OT VII, the highest level in Scientology, to Advanced Organizations around the world. [3]

November 1970

Douglas Caddy leaves the Mullen firm to work for Gall, Lane, Powell and Kilcullen, where tax and probate attorney Meade Emory is connected.

E. Howard Hunt becomes a "client" of Caddy and of Gall, Lane, Powell and Kilcullen. Caddy consults with Hunt regarding probate and other matters. G. Gordon Liddy is approached by Robert Mardian, asking Liddy to take a position that Mardian describes as "super-confidential."[51]

1971

January 1971

NSA's Harold "Hal" Puthoff, one of fewer than 3,000 Scientology "Clears" in the world in 1971, has joined the ranks of a much smaller number of OT VIIs.

NSA's Hal Puthoff somehow has gotten past L. Ron Hubbard's prohibitions against government spy agency personnel being allowed access to upper-level Scientology, and has progressed up the Scientology levels to the recently-released OT VII—the highest level available. He writes a success story for a Scientology publication about having completed OT VII, saying that on a weekend he had stood outside a locked building and remotely viewed information he wanted from a building directory that he couldn't physically read from the doorway, then verified later, when the building was open, that what he had viewed remotely had been accurate.

[NOTE: According to Scientology's the Auditor magazine, Special Issue March 1971, by that date there are only 2,773 Scientology "Clears" in the world. Being Clear is a prerequisite to the OT Levels. Even if 10% had gone all the way up through the higher levels by this date, Puthoff would be among only about 300 OT VIIs in the world.] [52]

February 1971

A hidden taping system is installed in the Oval Office of the White House. [22]

April 1971

Vietnam vet and Green Beret Paul Preston has signed up in Scientology's Sea Org, and is at the confidential land base for Scientology—"Tours Reception Center," in Tangier, Morocco—where the Flagship Apollo often docks with L. Ron Hubbard aboard.

Saturday, 17 April 1971

CIA's Bernard Barker

E. Howard Hunt is in Miami and meets with Bernard Barker, Eugenio Martinez, and Felipe De Diego. Bernard Barker has a history of almost seven years with CIA. Eugenio Martinez is on "retainer" with CIA.

[NOTE: A little over four months later, these same three men will be involved with Hunt in a purported break-in of the offices of psychiatrist Lewis Fielding, ostensibly in response to Daniel Ellsberg having leaked the Pentagon Papers. But the Pentagon Papers haven't been leaked to the press yet, and won't be for almost two months.] [1]

May 1971

IRS begins an audit of the senior Scientology corporation, Church of Scientology of California. Meade Emory, who later will corporately restructure all of Scientology, is Legislative Attorney for the Joint Committee on Taxation. [28]

June 1971

Daniel Ellsberg makes "a series of phone calls" to psychiatrist Lewis Fielding "shortly before" the Pentagon Papers are published. [2]

Saturday, 12 June 1971

The day before the "Pentagon Papers" are published, Morton Halperin, Leslie Gelb, and Defense Department official Paul Nitze make "a deposit into the National Archives" of "a whole lot of papers."

[NOTE: This turns out later to be copies of the not-yet-published Pentagon Papers that will make Daniel Ellsberg famous and launch everything that later comes to be known as "Watergate."] [22]

Sunday, 13 June 1971

Daniel Ellsberg, having highest possible clearances from CIA, leaks the "Pentagon Papers"

The New York Times publishes the first of three installments of secret documents that have been passed to Times reporter Neil Sheehan by Daniel Ellsberg. These come to be known as the "Pentagon Papers."

Tuesday, 15 June 1971

G. Gordon Liddy is abruptly transferred from being "Special Assistant to the Secretary of the Treasury" to "Staff Assistant of the President of the United States," part of the White House Domestic Council. Liddy is supplied with White House credentials. [53] [51]

Wednesday, 16 June 1971

G. Gordon Liddy is at the White House in his new job. A William Galbraith comes to the White House, purportedly one of a group of officers from the White House News Photographers Association (WHNPA).

[NOTE: No "William Galbraith" has been found on WHNPA rolls, and a William Galbraith has been identified as having been a CIA agent. Nine days after this event a girl is found shot to death on board the Scientology Flagship "Apollo" at Safi, Morocco, and a William Galbraith will be in Safi representing the U.S. Embassy in Morocco in response. See 25 June 1971 and 13 July 1971.] [51] [54]

Friday, 18 June 1971

An unspecified amount of money being carried by John McLean while on a confidential mission from the Scientology Flagship Apollo is reported stolen by McLean. Also with the Apollo at the time is Green Beret Paul Preston.

[NOTE: McLean will later become a key witness for the Commissioner of Internal Revenue against Scientology in the IRS case that results from a tax audit of Scientology being pursued at the time this event takes place. See May 1971.] [28]

Friday, 25 June 1971

One week to the day after John McLean reports Scientology money having been stolen from him, Susan Meister is found shot to death in a cabin aboard the Flagship Apollo, docked in Safi, Morocco. Conflicting reports say she was shot either in the mouth or in the forehead. One report says the gun was folded in her hands neatly in her lap. Her death is ruled a suicide by Moroccan authorities. [46] [55] [56]

Monday, 28 June 1971

Daniel Ellsberg is indicted for the leak of the Pentagon Papers.

Wednesday, 30 June 1971

Yvonne Gillham at Scientology's Celebrity Centre is closely connected to both Hal Puthoff and Ingo Swann

The Supreme Court rules 6-3 that the government has not shown compelling evidence to justify blocking further publication of the Pentagon Papers.

July 1971

Yvonne Gillham, Executive Director of Scientology's Celebrity Centre in Los Angeles, is in regular touch with both Ingo Swann and Hal

Puthoff. As Executive Director, as a fellow OT, and as a highly-trained Scientology auditor, she has taken on responsibility for their progress in and connection to Scientology.

Thursday, 1 July 1971

David Young—who is with NSA, the same agency as OT VII Hal Puthoff—is appointed to the White House Domestic Council to work with Egil Krogh [57] [2]

On or about the same date, Carol Ellsberg, Daniel Ellsberg's ex-wife, calls the FBI. She tells them that Daniel Ellsberg had seen a psychiatrist. She says that Ellsberg has "assured her" that he "had told this analyst all about what he had done" (referring to the Pentagon Papers). She volunteers the name of the Beverly Hills psychiatrist: Lewis Fielding.

[NOTE: Daniel and Carol Ellsberg have been living apart since January 1964, divorced since 1966. Daniel Ellsberg didn't begin with Fielding until two years after the divorce, in March of 1968 (see), and had quit seeing Fielding in September 1970 (see)—nearly a year before "what he had done."]

Jack Caulfield

On or about the same date, John "Jack" Caulfield, Staff Assistant to President Nixon, has created a 12-page political espionage proposal called "Sandwedge." Ostensibly as part of it, Anthony Ulasewicz has rented an apartment at 321 East 48th Street (Apartment 11-C), New York City. G. Gordon Liddy is given the complete "Sandwedge" plan.

[NOTE: The apartment is in close proximity to the lab and school of CIA's Cleve Backster. It provides a backstopped New York address and phone. Note, too, that the reference for date of Sandwedge is a document in the National Archives titled "7/71 Sandwedge proposal," despite most anecdotal accounts placing it later in 1971.] [58] [59]

Friday, 2 July 1971

CIA Director Richard Helms is pushing behind the scenes to get E. Howard Hunt into a position connected with the White House in response to the Pentagon Papers having been leaked. H. R. Haldeman tells Nixon that that Helms has described Hunt: "Ruthless, quiet and careful, low profile. He gets things done. He will work well with all of us. He's very concerned about the health of the administration. His concern, he thinks, is they're out to get us and all that, but he's not a fanatic. We could be absolutely certain it'll involve secrecy...."

Charles Colson sends a memo to H. R. Haldeman with a transcript of a phone conversation he had with E. Howard Hunt the previous day—which he happened to record. Colson says: "The more I think about Howard Hunt's background, politics, disposition and experience, the more I think it would be worth your time to meet him."[22] [1]

Wednesday, 7 July 1971

E. Howard Hunt is hired as a "White House consultant" while keeping his full-time job at Mullen. Hunt is supplied with White House credentials. [1]

Thursday, 8 July 1971

E. Howard Hunt has a private meeting with CIA's Lucien Conein, Hunt's acquaintance of almost 30 years.

[NOTE: Conein had been part of the team that Daniel Ellsberg had gone with to Vietnam, headed by CIA's Edward Lansdale [often misspelled as Edward Landsdale], in 1965-66.] [1]

Tuesday, 13 July 1971

A William Galbraith, represented as being American vice consul from Casablanca, meets in Safi, Morocco with two representatives from the Scientology Flagship Apollo, ostensibly concerning the 25 June 1971 death of Susan Meister [see 16 June 1971 re: Galbraith].

A William Galbraith, reportedly CIA, is at the Scientology Flagship Apollo in Safi, Morocco after having been at the White House a month before, ostensibly as a White House news photographer.

Tuesday, 20 July 1971

E. Howard Hunt has a private meeting with CIA's Edward G. Lansdale [often misspelled as Edward G. Landsdale].

[NOTE: Lansdale had taken Daniel Ellsberg and Lucien Conein to Vietnam in 1965-66. Green Beret Paul Preston had also been there as part of so-called "pacification" programs," as well as Neil Sheehan, who just has published the Pentagon Papers.][1]

Thursday, 22 July 1971

E. Howard Hunt goes to CIA headquarters and meets privately with Deputy Director of CIA Robert Cushman. [1] [60]

Friday, 23 July 1971

The CIA supplies E. Howard Hunt with counterfeit ID in the name of "Edward J. Warren." Hunt meets CIA's Stephen Greenwood in a CIA safehouse where a fake driver's license and other ID material, plus a disguise, are given to Hunt. [60] [1] [61]

Saturday, 24 July 1971

NSA's David Young is running everything that leads to the Fielding office break-in. Young will later be given immunity by Watergate prosecutors, then will report the Fielding burglary, backed up by CIA photos, just after CIA has given a secret contract to Hal Puthoff to develop the remote viewing program using OT VII Ingo Swann.

Based on a memorandum by Egil Krogh and NSA's David Young, the Special Investigations Unit is established at the White House under them. It comes to be known as the White House Plumbers.

[NOTE: David Young gives the unit its nickname, supposedly because it is there to "stop leaks." It never stops a single leak, or accomplishes anything effective regarding security leaks. Liddy and Hunt are already established in their positions weeks before the unit is created. The creation of the Special Investigations Unit does nothing to alter the operational status or position of either of them.]

Friday, 30 July 1971

A highly secure facility has been set up in Room 16 of the Old Executive Office Building adjacent to the White House that G. Gordon Liddy and E. Howard Hunt use. It includes a secure phone used "mostly to talk to the CIA at Langley."[51]

Early August 1971

Green Beret and Vietnam veteran Paul Preston is aboard the Flagship Apollo in the Mediterranean, where L. Ron Hubbard is living.

[NOTE: Less than a year later, Hubbard disappears, and some sources say Preston is with Hubbard at the time of the disappearance as a "bodyguard." See 28 May 1972.]

G. Gordon Liddy is in regular communication with "State and the CIA," having direct conversations with CIA Director Richard Helms. Liddy is briefed by CIA on "several additional sensitive programs in connection with his assignment to the White House staff." Liddy is also making regular trips to the Pentagon.

E. Howard Hunt is making regular trips to the State Department. U.S. Ambassador to the United Nations at the time is George H.W. Bush (Sr.). [62] [51] [1] [34]

Monday, 2 August 1971

CIA psychiatrist Bernard Malloy comes to Room 16 and meets privately with G. Gordon Liddy and E. Howard Hunt. [1] [53]

Friday, 6 August 1971

E. Howard Hunt again meets clandestinely in a CIA safehouse, this time with CIA's Stephen Greenwood and also with CIA's Cleo Gephart. Hunt purportedly discusses CIA providing a "backstopped address and phone" in New York city. Hunt also asks for CIA to provide phony ID and a disguise for "an associate"—G. Gordon Liddy.

[NOTE: Hunt is asking for ID and disguise for Liddy prior to any proposal to break into Lewis Fielding's office. Also, there's already a backstopped address and phone in New York city at 321 East 48th Street, Apartment 11-C, New York City, set up by Anthony Ulasewicz as part of the Sandwedge proposal, which Liddy and Hunt have. See 1 July 1971.] [61] by the CIA, and both have been issued White House credentials.

Wednesday, 11 August 1971

CIA psychiatrist Bernard Malloy again comes to Room 16 and meets privately with G. Gordon Liddy and E. Howard Hunt. Soon after, Liddy and Hunt recommend an attempt at surreptitious entry for "acquisition of psychiatric materials" on Daniel Ellsberg from the files of psychiatrist Lewis Fielding. They claim the need, first, for a "feasibility study" of Fielding's Beverly Hills office. [1] [53]

Thursday, 12 August 1971

L. Ron Hubbard issues a policy letter called "Advanced Courses" that makes access to all the confidential upper levels of Scientology available by invitation only, to be based largely on ethics record in Scientology and security of materials. [63]

Friday, 20 August 1971

The CIA supplies G. Gordon Liddy with counterfeit ID in the name of "George F. Leonard." Hunt and Liddy meet CIA's Stephen Greenwood (called "Steve" in Hunt's account)

Hunt and Liddy take photographs of each other in front of Fielding's door in CIA-supplied "disguises." The photos will later be used by CIA to give Ellsberg a convenient "Get Out of Jail Free" card.

in a CIA safehouse where a CIA-created fake driver's license and other ID material, plus a disguise, and a camera are issued to Liddy.

[NOTE: According to Greenwood, Hunt and Liddy say they have to "stop by the Pentagon" on their way to the airport, although they don't say where they are going. It isn't to Los Angeles for the Fielding office "feasibility study," since that doesn't take place until 26 August 1971 (see) according to the available accounts from Hunt and Liddy, cited in this timeline.] [60] [1] [61]

Thursday, 26 August 1971

E. Howard Hunt and G. Gordon Liddy fly to Los Angeles. Hunt takes pictures of Liddy, in his CIA-issued black wig, standing in front of psychiatrist Lewis Fielding's office door, with Fielding's name on the door. Liddy also takes pictures of Hunt. The photos are taken with the camera supplied to them by CIA. [51] [1] [64]

Friday, 27 August 1971

E. Howard Hunt and G. Gordon Liddy fly back to Washington, D.C. CIA's Stephen Greenwood meets them at the airport, where Hunt gives Greenwood the film for developing by CIA. Greenwood delivers prints to Hunt the same day. The CIA keeps a copy of the photos of Liddy and Hunt (in CIA-provided "disguises" that don't disguise them at all) mugging in front of Lewis Fielding's identifiable door.

[NOTE: The CIA later turns their copies of the photos over to Watergate investigators, which results in all criminal charges against Daniel Ellsberg for leaking the Pentagon Papers to be dropped. See 1973, specifically 3 January, 17 March, 15 April, and 11 May 1973] [60] [1] [61]

Saturday, 28 August 1971

CIA's Eugenio Martinez

On a Saturday, Hunt and Liddy purportedly are in Room 16 when Liddy tells Hunt that the plan to do a break-in of Fielding's office is approved, but that the two of them are not "to be permitted anywhere near the target premises." [See 27 August 1971, immediately above.] E. Howard Hunt then purportedly calls Bernard Barker in Miami and asks if Barker can "put together a three-man entry team." Barker calls back to say it will be Barker, Eugenio Martinez, and Felipe De Diego.

[NOTE: As luck would have it, this happens to be the same three men Hunt had met with in Miami two months before the Pentagon Papers were published. See 17 April 1971.] [1] [51] [65]

Friday, 3 September 1971

A break-in takes place at the office of psychiatrist Lewis J. Fielding in Beverly Hills, California. The break-in is made obvious by the smashing of a window. Accounts of the break-in are irreconcilably conflicting. According to Bernard Barker, E. Howard Hunt, and G. Gordon Liddy, the three Cubans—Barker, Martinez, and De Diego—had entered the office and searched thoroughly, and there was no file on Daniel Ellsberg anywhere. According to Lewis Fielding, there was a file on Ellsberg in his office, which Fielding says he found on the floor the next morning. Fielding claims it was evident that someone had gone through the file.

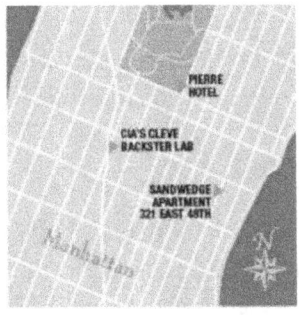

Liddy and Hunt in New York on same night as Fielding break-in in Los Angeles

The same night, Hunt and Liddy are in New York City—where Hunt has made an issue of needing "a backstopped address." They check into the Pierre hotel and remain in New York through at least Sunday, 5 September 1971.

[NOTE: There is no physical evidence that either Liddy or Hunt had been in Los Angeles at all for the Fielding office break-in. Only the anecdotal claims of the co-conspirators account for the whereabouts of Hunt and Liddy that weekend. This is similar to the later purported Watergate first break-in that involved the same personnel. (See 26, 27, and 28 May 1972.) Also, there is a backstopped address that was available to Liddy and Hunt in New York: 321 East 48th Street, not far from the Pierre hotel. Both locations are less than a mile from the Times Square lab and polygraph school of CIA's Cleve Backster. Five days after Hunt and Liddy leave New York (see 9 September 1971), OT VII Ingo Swann will "chance" to meet Backster "at a party" in New York. Backster will be the person who then ostensibly sets up a connection between Ingo Swann—a Scientology OT VII—and NSA's Hal Puthoff, also a Scientology OT VII. Both will be contracted by CIA to start CIA's secret remote viewing program (see 1 October 1972).] [1] [51] [65]

Thursday, 9 September 1971

OT VII Ingo Swann meets CIA's Cleve Backster, purportedly "at a party" in New York. Backster has an "extensive network of contacts in law enforcement agencies and within the CIA."[44]

Sunday, 12 September 1971

OT VII Ingo Swann visits Cleve Backster's lab and polygraph school in New York city where Swann is asked to think thoughts of harming a plant that Backster has connected up to what Swann says was "a polygraph." Swann thinks of lighting a match with the intent of burning one of the plant's leaves, and there is an immediate and violent reaction. With repetitions, the reaction diminishes, and the conclusion

OT VII Ingo Swann is drawn that not only is the plant capable of detecting harmful thought, but can "learn" to differentiate between true and artificial intent. The thought directed at the plant is changed to one of putting acid in its pot, with the same curve of results. [44]

Wednesday, 15 September 1971

A "Master File" of cables (telexes) disappears from the External Comm Bureau of the Scientology Flagship Apollo. The file contains all cables related to the administration of Scientology worldwide from 22 August 1971 to 15 September 1971. In the External Comm Bureau at the time is John McLean. Also, aboard his ship is Green Beret Paul Preston, doing a service called "Word Clearing Method 1."[62]

E. Howard Hunt makes a request that the CIA "immediately recall a 24-year-old secretary" from Paris for his use and "explain to all concerned that she was urgently needed for an unspecified special assignment."[2]

Monday, 20 c. September 1971

E. Howard Hunt is granted special permission by the State Department for "full access to the department's chronological cable files."

[NOTE: Shortly thereafter, Hunt is engaged in forging cables.] [1]

Saturday, 25 September 1971

Scientology OT VII Ingo Swann is doing experiments with Cleve Backster involving a piece of graphite hooked into a Wheatstone bridge [the main mechanism in a Scientology e-meter], connected with a chart recorder. Swann learns that he can focus a "beam" of intention at the graphite, and cause repeatable jogs in the chart. [44]

October 1971

E. Howard Hunt is in telephone contact with CIA Chief European Division John Hart, and has several telephone conversations with CIA Executive Officer European Division John Caswell. [34]

Friday, 1 October 1971

APA's Karlis Osis

Dr. John Wingate of the American Society for Psychical Research (ASPR) in New York invites OT VII Ingo Swann to work on "well-designed" psychic experiments with psychiatrist Karlis Osis. Osis is a member of the very anti-Scientology American Psychological Association (APA).

[NOTE: Chicago's W. Clement Stone is a member and major contributor to the ASPR. About 7 months later, CIA's E. Howard Hunt will deliver an undisclosed amount of cash in a sealed envelope to the W. Clement and Jessie V. Stone Foundation in Chicago. See April 1972.] [44]

Sunday, 10 October 1971

Cleve Backster writes and circulates a small report entitled "Psychokinetic Effects on Small Samples of Graphite," detailing the repeatable experiments that he has conducted with OT VII Ingo Swann. Backster tells Swann, "Boy, are the guys down at the CIA going to be interested in you."[44]

Friday, 15 October 1971

E. Howard Hunt meets with CIA Director Richard Helms. Ingo Swann meets with Gertrude Schmeidler, who asks Swann if he thinks he can influence a thermistor isolated in a sealed thermos bottle. He agrees to try. [44] [38]

Late October 1971

Scientology OT VII Ingo Swann is in Washington, D.C. with "a colleague" meeting "in bars and pizza parlours" with unnamed intelligence personnel. At one of the meetings with "six spooks," Swan is asked: "If you were going to set up a threat analysis program to match what the Soviets are up to, what would you do?"'[66]

By late October 1971 Ingo Swann is in Washington, D.C. meeting with U.S. intelligence agency personnel.

Saturday, 30 October 1971

Ingo Swann is working with CIA's Cleve Backster, testing "psi probes" on gasses in pressurized containers. He and Backster move on to experiments with biologicals, including one-celled biological specimens, blood, and seminal fluid. When Swann has some success in affecting biologicals with psychic probing, Backster says, "Well, you've just done something the Soviets have been working on for a long time."[44]

Early November 1971

CIA's James McCord, purportedly retired in August 1970, signs a contract with the Republican National Committee to handle "security." The contract is in the name of "McCord Associates, Inc."

[NOTE: The corporation will not be created until several weeks after the contract is signed; incorporation papers are not filed until 19 November 1971 (see) in Maryland.] [67]

Monday, 15 November 1971

Gertrude Schmeidler

Dr. Gertrude Schmeidler conducts her thermistor experiments with OT VII Ingo Swann at the New York City College. Swann can produce changes in the target thermistors, while the control thermistors remained unchanged, on a repeatable basis at the direction of the experimenter. [44]

Friday, 19 November 1971

CIA's E. Howard Hunt contacts CIA's Office of Security Director Robert Osborne. [38]

CIA's James McCord files incorporation papers in Maryland for McCord Associates, Inc., ostensibly a security company, but the incorporation papers say nothing about providing security, and the company is not licensed for security. Included on the board are McCord, his wife, and his sister, Dorothy Berry, who works for an "oil company in Houston."

[NOTE: Berry later claimed she had "no idea" she had been listed on the board. Also, the Gulf Resources and Chemical Corporation—an "oil company in Houston" that controls half the world's supply of lithium—will later provide checks that get converted to traceable $100 bills for part of what becomes known as Watergate. See 15 April 1972.] [67]

Monday, 22 November 1971

OT VII Ingo Swann is involved in one of a series of ten out-of-body (OOB) perception experiments at ASPR, the task being to verbally describe objects out of his sight in a target tray. Having difficulty doing a narrative description of the target items, he hits upon the idea of doing sketches. Successful, this becomes a regular part of the experiments. [44]

Wednesday, 1 December 1971

Gertrude Schmeidler's paper, "PK Effects Upon Continuously Recorded Temperature," describes results of her thermistor experiments with Ingo Swann and is being circulated for peer review. It generates speculation that if someone could trigger a thermistor, they also might be able to remotely trigger a bomb. There are requests for interviews of Schmeidler and Swann from media like Time and Newsweek, but Swann refuses. [44]

Monday, 6 December 1971

G. Gordon Liddy becomes General Counsel to the Committee for the Re-Election of the President. [68] [51]

Wednesday, 8 December 1971

In one of a series of long-distance remote viewing experiments at ASPR, OT VII Ingo Swann suggests calling the experiments "remote sensing" or "remote viewing."[44]

E. Howard Hunt is in touch with senior CIA officer Peter Jessup, who is with the National Security Council staff. [38]

On or about the same day, Hunt meets privately again with CIA's Lucien Conein.[1]

Sunday, 12 December 1971

NSA's David Young meets with Egil Krogh and CIA psychiatrist Bernard Malloy. [2]

Lt. George W. Bush

Thursday, 16 December 1971

CIA's E. Howard Hunt is in Dallas, Texas—an airline hub.

Lt. George W. Bush is living in Houston, Texas. He is a pilot trained on T-38 Talons, a type of plane used as a chase plane. [69]

Thursday, 30 December 1971

An "out-of-body" (OOB) perception experiment results in Ingo Swann sketching the target object (a 7-Up can) upside down, so he believes he has missed getting it. Dr. Osis realizes that it is a perfect drawing of the can once it is turned upside down. [44]

1972

January 1972

G. Gordon Liddy and E. Howard Hunt are collaborating on a "political espionage" plan to replace the Sandwedge proposal. One of the items they have factored into the budget, ostensibly for "political espionage," is a chase plane. [1] [51]

Liddy and Hunt in New York on same night as Fielding break-in in Los Angeles

T-38 Talon, commonly used as "chase plane"

Monday, 10 January 1972

G. Gordon Liddy is in New York city at the apartment Ulasewicz has established at 321 East 48th Street, Apartment 11-C. [70] [51]

Wednesday, 12 January 1972

G. Gordon Liddy is still in New York city. Ingo Swann learns that "two men in suits," flashing credentials, have visited the ASPR facility investigating him. They have met with Dr. Osis, and have looked at the experiment rooms and some of the experiment results. Osis tells Swann that he (Osis) isn't "free to tell" Swann what was discussed. [44]

Friday, 14 January 1972

Origin date of Defense Intelligence Agency (DIA) report, "Controlled Offensive Behavior—USSR"[71]

Tuesday, 25 January 1972

Buell Mullen tells Ingo Swann that a small group of her "high-placed friends" has begun establishing a pool of money for Swann. Already some $70,000 has been "pledged" from "several sources."[44]

Monday, 31 January 1972

"Information Cut-off Date" for a 1972 Defense Intelligence Agency (DIA) report entitled "Controlled Offensive Behavior—USSR" concerning Soviet research and development of "psi" technologies. [71]

E. Howard Hunt and G. Gordon Liddy return from a weekend trip to Los Angeles, during which Liddy also has gone to San Diego and back.

[NOTE: Dr. Augustus B. Kinzel has a home just outside San Diego.] [1] [51]

The secret DIA report, "Controlled Offensive Behavior—USSR" will say when published that Soviet knowledge in parapsychology "is superior to that of the U.S."

Early February 1972

Buell Mullen calls Ingo Swann to say that Dr. Augustus B. Kinzel will be in New York city on 17 February with "some friends" who want to meet with Swann. She is having a dinner party for the occasion. Swann says he'll be there. [44]

G. Gordon Liddy and E. Howard Hunt fly to Miami, home of Bernard Barker and other CIA-connected Cubans. [1] [51]

G. Gordon Liddy "recruits" CIA's James McCord as a "wire man," purportedly to be able to do electronic eavesdropping for "political espionage" purposes.

[NOTE: At the time, Liddy has no approved budget for any such activities, nor are there any approved plans for, or targets for, any such activities.] [53]

Thursday, 17 February 1972

Guess Who's Coming to Dinner

At a dinner at Buell Mullen's home in New York, Dr. Augustus B. Kinzel has brought four "friends" in suits who Kinzel will only introduce to Swann by first names. They have a one-hour meeting that is "strictly confidential," concerning "big-time funding for a new research organization" that's separate from the $70,000 already collected. According to Swann, at least one of the "friends" is CIA. [44]

On or about the same date, E. Howard Hunt and G. Gordon Liddy again fly to Miami, ostensibly to meet with Donald Segretti (a.k.a.

"Donald Simmons"). While there, Hunt is in contact with CIA's Bernard Barker.

Tuesday, 22 February 1972

OT VII Ingo Swann performs the first of a series of "out-of-body" (OOB) experiment with Vera Feldman of the ASPR as the outbound experimenter. Swann is hooked up to brainwave leads and locked in the OOB room while Feldman goes to the Museum of Natural History a few blocks away. Swann gets a high number of "hits" on what Feldman is seeing, one of them being a display case full of gemstones. Swann and Feldman talk about ESP being used for psychic spying. [44]

Liddy and Hunt name the operations they are engaged in "GEMSTONE"

G. Gordon Liddy meets with CIA in connection with CIA "special clearances" he has been granted. [34]

Thursday, 24 February 1972

G. Gordon Liddy and E. Howard Hunt meet with a "retired" CIA doctor, introduced by Hunt to Liddy as "Dr. Edward Gunn," to get briefed by him on various covert means of murder for a possible assassination.

[NOTE: Although Liddy and Hunt relate many similar incidents, if disjointedly, in their respective autobiographies, Hunt mentions nothing about this incident in his, while Liddy devotes several pages to it.]

Late February 1972

OT VII Ingo Swann, connected with ASPR, meets Robert D. Ericsson, Executive Director of Spiritual Frontiers Fellowship (SFF).

[NOTE: Chicago's W. Clement Stone is a member and major contributor to both the SFF and ASPR. About two months later, E. Howard Hunt will deliver an undisclosed amount of cash in a sealed envelope to the W. Clement and Jessie V. Stone Foundation in Chicago. See April 1972.] [44]

One of the members of the board of ASPR, A.C. Twitchell, Jr., purportedly calls Ingo Swann early in the morning saying that there is a move afoot at the ASPR to have Swann ejected on the grounds that he is a Scientologist. Twitchell says that it has been circulating that Swann is "Hubbard's spy," and is seeking to take over the ASPR on Hubbard's behalf. Swann threatens to sue the board over his civil rights. [44]

E. Howard Hunt travels to Nicaragua on an "undisclosed mission."

[NOTE: See entry for 3 March 1972.] [38]

Wednesday, 1 March 1972

Russell Targa, Charles T. Tart, and David Hart release a proposal entitled "Research on Techniques to Enhance Extra Sensory Perception."[44]

On or about the same date, Douglas Caddy begins to do "legal tasks" for G. Gordon Liddy. [72]

Friday, 3 March 1972

Gary O. Morris, psychiatrist of E. Howard Hunt's wife, Dorothy, vanishes while on vacation on the Caribbean island of St. Lucia. No trace is ever found of the pleasure boat he had left on for a cruise with his wife and a local captain, Mervin Augustin. [38]

Wednesday, 15 March 1972

A memorandum is sent to Director of FBI J. Edgar Hoover from the Legal Attaché (LEGAT) Copenhagen titled "SUBJECT: L. RON HUBBARD." It says: "On 3/13/72, [BLACKED OUT] advised that he has not yet completed preparation of his report concerning the Scientology Organization and its operations in Denmark. He reiterated, however, that when completed a copy of this report will be designated for [BLACKED OUT] Contact will be maintained with [BLACKED OUT] in order to ensure that this office receives copies of his report and Bureau will be kept advised."[73]

Monday, 20 March 1972

OT VII Ingo Swann is at Cleve Backster's lab in New York. Backster hands some papers to Swann on Hal Puthoff and purportedly says, "You two might get along. He's into Scientology, too."

[NOTE: Both Hal Puthoff and Ingo Swann have been connected with Yvonne Gillham at Celebrity Centre for some time. Both also are OT VII, and the only place in the United States delivering the OT Levels is the Advanced Organization in Los Angeles (AOLA), where Yvonne Gillham had been the senior executive before starting Celebrity Centre.] [44]

Wednesday, 22 March 1972

Janet Mitchell writes regarding out-of-body brightness comparison experiments with Ingo Swann, saying, "It may be possible that he can see all the waves in the atmosphere from infrared to ultraviolet."[44]

Monday, 27 March 1972

G. Gordon Liddy's job abruptly changes to general counsel of the Finance Committee to Re-elect the President. [53]

Wednesday, 29 March 1972

Two days after Liddy's job changes, E. Howard Hunt "terminates" in his paid capacity as a White House consultant—yet he keeps his office and the safe he'd used as such, and keeps his White House credentials because he continues to "work there a few hours each week."[22] [1]

Thursday, 30 March 1972

The day after E. Howard Hunt's "official" disconnection from the White House, OT VII Ingo Swann contacts OT VII Hal Puthoff saying Cleve Backster has "suggested" for Swann to contact Puthoff. Swann has several phone conversations over several days with Puthoff, who suggests that Swann come out to Stanford Research Institute (SRI) for a couple of weeks to do some experiments. [44]

Early April 1972

Unknown amount of cash delivered by CIA's E. Howard Hunt to offices of W. Clement Stone's foundation

CIA's E. Howard Hunt flies to Chicago and delivers an undisclosed amount of cash in a sealed envelope to W. Clement and Jessie V. Stone Foundation. [1]

Tuesday, 4 April 1972

OT VII Ingo Swann receives word that an independent judge, blind to the fact that she was judging an experiment in out-of-body (OOB) perceptions, has correctly matched all eight of the former "picture drawing" trials—a 100 per cent match between the OOB drawings and the contents of the target trays. [44]

Friday, 7 April 1972

L. Ron Hubbard gives three taped lectures to students on the Expanded Dianetics course. They are the last public lectures Hubbard ever will give.

[NOTE: As of this date, L. Ron Hubbard had given over 1,300 public lectures since 1950—averaging a little over one a week.]

Monday, 10 April 1972

A timely Minnesota court ruling puts a shipment of Scientology E-meters into permanent federal custody and control

A court ruling this date by the United States Court of Appeals, Eighth Circuit, in St. Paul, Minnesota, allows the U.S. federal government to keep a shipment of Scientology e-meters that had been seized by the Department of Health, Education and Welfare on the basis of "improper labeling," putting an unknown number of e-meters in permanent custody and possession of federal agencies. [74]

Saturday, 15 April 1972

E. Howard Hunt and G. Gordon Liddy fly to Miami and deliver checks drawn on a Mexico City bank to CIA's Bernard Barker.

[NOTE: Several of the checks have originated from Gulf Resources and Chemical Corporation in Houston, which at the time controls half the world's supply of lithium, used in the making of hydrogen bombs and in psychiatric drugs.] [1]

Monday, 17 April 1972

Physicist Dr. Russell Targ meets with CIA personnel from the Office of Strategic Intelligence (OSI) and discusses the subject of paranormal abilities. Films of Soviets moving inanimate objects by mental powers are made available to analysts from OSI. [19]

Thursday, 20 April 1972

CIA Office of Strategic Intelligence personnel who have been briefed by Russell Targ contact personnel from the Office of Research and Development (ORD) and Technical Services Division (TSD) regarding films and reports of Soviet investigations into psychokinesis.

[NOTE: Although the name of CIA's Technical Services Department (TSD) later changes to Office of Technical Services (OTS), some sources anachronistically refer to TSD as OTS when it was still TSD. The name isn't officially changed until November 1972.] [19]

Monday, 24 April 1972

CIA's Bernard Barker cashes a cashier's check for $25,000 at his bank in Miami.

[NOTE: This $25,000, from the Dahlberg check, plus two later withdrawals by Barker will equal $114,000. See 2 May and 8 May 1972.] [75] [76]

Monday, 1 May 1972

Russell Targ has joined the Stanford Research Institute, and is visited by a CIA Office of Research and Development (ORD) Project Officer. Targ proposes that some psychokinetic verification investigations can be done at SRI in conjunction with Scientology OT VII Hal Puthoff. [19]

J. Edgar Hoover found dead

CIA's James McCord contacts an ex-FBI agent, Alfred Baldwin, who is living in Connecticut. McCord purportedly doesn't know Baldwin, but wants Baldwin to come to Washington, D.C. that night. [77]

Tuesday, 2 May 1972

FBI Director J. Edgar Hoover is found dead in his home in the early morning hours. L. Patrick Gray—who has no background in law enforcement—is appointed as Acting Director of FBI.

[NOTE: Hoover's death is attributed to a heart attack, and no autopsy is done. L. Patrick Gray later will destroy material taken from the White House safe of E. Howard Hunt, then will resign.]

CIA's Bernard Barker withdraws an unspecified amount of cash from his bank in Miami.

[NOTE: This is the second of three transactions by Barker that will total $114,000.] [75]

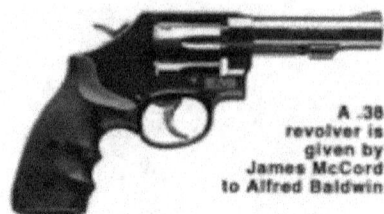

Alfred Baldwin meets with James McCord. McCord issues Baldwin a Smith & Wesson .38 snub-nose revolver. Baldwin is assigned to travel as a bodyguard with Martha Mitchell on "a trip to the Midwest."[77]

Wednesday, 3 May 1972

OT VII Ingo Swann performs an experiment that he says "scared the bejesus out of the experimenters, and parapsychology as well." In it, Swann perceives not just things the two outbound "beacons" are seeing, but also senses confusion in them. When they come back and confirm that they had gotten lost in some construction work being done in the Museum of Natural History, one says with concern, "Does this mean you can read our minds, too?"[44]

CIA's Bernard Barker, Eugenio Martinez, Frank Sturgis, and Filipe De Diego arrive in Washington, D.C. from Miami and meet with G. Gordon Liddy and E. Howard Hunt. [65] [1]

Thursday, 4 May 1972

Lt. George W. Bush is ordered to "report to commander, 111 F.I.S., Ellington AFB, not later than (NLT) 14 May, 1972."

[NOTE: Bush does not report as ordered. See 19 May 1972.] [78]

Friday, 5 May 1972

CIA's James McCord rents room 419 of the Howard Johnson's motel across the street from the Watergate. The room is registered in the name of McCord Associates. [79]

Monday, 8 May 1972

Alfred Baldwin returns to Washington, D.C. from his trip with Martha Mitchell. He is told by James McCord to keep the .38 revolver because "he might be going on another trip."[77]

G. Gordon Liddy, in D.C., calls CIA's Bernard Barker in Miami. [79]

Bernard Barker withdraws another unspecified amount of cash from his bank in Miami which, with two other transactions, now totals $114,000. [75]

Though not all in this timeline, cash pay-outs to CIA's James McCord will total at least $71,000

James McCord receives $4,000 in cash from G. Gordon Liddy. [80]

Tuesday, 9 May 1972

Alfred Baldwin leaves Washington, D.C., ostensibly going to his home in Connecticut to "get more clothes." He takes the .38 revolver with him, purportedly because he has been told by James McCord that he might be going on another trip with Martha Mitchell that is scheduled for 11 May 1972.

[NOTE: Baldwin doesn't return until 12 May 1972.] [77]

Wednesday, 10 May 1972

CIA's James McCord is in Rockville, Maryland. He pays $3,500 cash for a "device capable of receiving intercepted wire and oral communications."

[NOTE: Rockville, Maryland is about six miles from Laurel, Maryland. Five days later presidential candidate George Wallace will be shot in Laurel, Maryland by Arthur Bremer with a .38 caliber revolver. See 15 May 1972.]

Friday, 12 May 1972

Alfred Baldwin returns to Washington, D.C. James McCord tells Baldwin he won't be going with Martha Mitchell so he can "turn in his gun." Baldwin purportedly gives the .38 revolver to McCord. McCord tells Baldwin to move from the Roger Smith hotel, where Baldwin has been staying, into room 419 at the Howard Johnson's motel. [77]

Monday, 15 May 1972

Presidential candidate George Wallace is shot in Laurel, Maryland with a .38 revolver

Presidential candidate George Wallace is shot by Arthur Bremer in Laurel, Maryland, ending his presidential campaign and partially paralyzing him.

Wednesday, 17 May 1972

CIA's Bernard Barker makes two calls from Miami to G. Gordon Liddy, and two calls to CIA's E. Howard Hunt. [79]

A memorandum is sent to Acting Director of FBI L. Patrick Gray from the Legal Attaché (LEGAT) Madrid titled "SUBJECT: L. RON

HUBBARD FPC." It says: "Enclosed for information and completion of Bureau and Legat, Copenhagen files is one copy of a memorandum dated 4/26/72, received from the [BLACKED OUT]."[73]

Friday, 19 May 1972

Ambassador to UN George H.W. Bush will become CIA Director, then President

Lt. George W. Bush (Jr.) contacts a superior officer in the reserves to discuss "options of how Bush can get out of coming to drill from now through November." The memo recording the conversation says that Bush "is working on another campaign for his dad." The memo writer thinks Bush is "also talking to someone upstairs."

[NOTE: George H. W. Bush (Sr.) is U.S. Ambassador to the U.N. at this time.] [81]

President Richard M. Nixon, about to embark on an historic trip to the Soviet Union, writes the following in a letter to Henry Kissinger and Alexander Haig: "The performance in the psychological warfare field is nothing short of disgraceful. The mountain has labored for seven weeks and when it finally produced, it produced not much more than a mouse. Or to put it more honestly, it produced a rat. We finally have a program now under way but it totally lacks imagination and I have no confidence whatever that the bureaucracy will carry it out. I do not simply blame (Richard) Helms and the CIA. After all, they do not support my policies because they basically are for the most part Ivy League and Georgetown society oriented."[82]

E. Howard Hunt makes two calls to Bernard Barker in Miami. [79]

Saturday, 20 May 1972

Richard Nixon leaves Washington, D.C. on his trip to Austria, the Soviet Union, Iran, and Poland. He will not return until 1 June 1972. [83]

James McCord sends Alfred Baldwin to Andrews Air Force Base, where Nixon is leaving on Air Force One, purportedly because there might be demonstrations and McCord wants Baldwin to be there for more "surveillance activities."

[NOTE: The "reason" supplied by McCord in testimony for this trip by Baldwin is too thin to slice, particularly in light of the amount of security surrounding Nixon's departure. Besides Air Force One, there is a fleet of White House planes at Andrews for use by VIPs and various staff connected with the White House.]

On or about the same day, CIA's E. Howard Hunt flies to Miami and meets with Bernard Barker. [1]

Monday, 22 May 1972

CIA's Frank Sturgis

Richard Nixon arrives in Moscow and is toasting Soviet leaders at a dinner. [83]

The CIA "Cuban contingent" arrives in Washington, D.C. from Miami: Bernard Barker, Frank Sturgis, Eugenio Martinez, and Virgilio Gonzalez. They are in D.C. purportedly to carry out a "first break-in" on the following weekend of Democratic National Committee headquarters at the Watergate with G. Gordon Liddy, CIA's E. Howard Hunt, and CIA's James McCord.

[NOTE: There is no physical evidence that any such "first break-in" ever took place. For full coverage, see Watergate first break-in. Note also that while E. Howard Hunt claims that six Cubans arrived on 22 May 1972, the referenced criminal appeals court ruling names only four.] [1] [84]

Tuesday, 23 May 1972

A memorandum is sent to Acting Director of FBI L. Patrick Gray from the Legal Attaché (LEGAT) Copenhagen titled "SUBJECT: L. RON HUBBARD FPC." It says: "Mr. Victor Wolf, Jr., U.S. Consul, American Embassy, Copenhagen, advised on 5/15/72 that he has not yet found time to prepare the report referred to in relet concerning the Scientology Organization. Mr. Wolf stated that he hopes to devote attention to this matter in a short time. This case will be placed in Pending Inactive status for a period of 90 days."[73]

Alfred Baldwin leaves Washington, D.C. again, purportedly going to his home in Connecticut again. No reason is given for his departure. [77]

CIA's Virgilio Gonzalez

Friday, 26 May 1972

G. Gordon Liddy, Alfred Baldwin, CIA's E. Howard Hunt, CIA's James McCord, and several Cuban CIA contract agents purportedly are engaged in a failed attempt to break into the Watergate—the "Americas dinner" attempt. [NOTE: But see Watergate first break-in.]

Saturday, 27 May 1972

G. Gordon Liddy, Alfred Baldwin, CIA's E. Howard Hunt, CIA's James McCord, and several Cuban CIA contract agents purportedly are engaged in a second failed attempt to break into the Watergate. [NOTE: But see Watergate first break-in.]

Sunday, 28 May 1972

There apparently was no "first break-in" at the Watergate. Then where were Liddy, Hunt, McCord, and Baldwin over Memorial Day weekend? AWOL with Lt. Bush?

G. Gordon Liddy, Alfred Baldwin, CIA's E. Howard Hunt, CIA's James McCord, and several Cuban CIA contract agents purportedly are engaged in a successful "first break-in" at DNC headquarters at the Watergate. According to their later claims, McCord placed two electronic bugs in the DNC headquarters during the "first break-in," and Bernard Barker purportedly had photos taken of the office of the Chairman, Lawrence O'Brien, and of documents on his desk.

[NOTE: There is no physical evidence that any such "first break-in" ever took place, or the purported two earlier failed attempts on the same holiday weekend. Barker later testified that he never was in O'Brien's office at all, and a telephone company sweep found no electronic bugs in the DNC at all (see 15 June 1972). For full coverage, see Watergate first break-in. There is nothing to account for the whereabouts of Liddy, Hunt, McCord, and Baldwin over the entire Memorial Day Weekend except the conflicting and contradictory anecdotal accounts of the co-conspirators themselves, which they volunteered when "caught" inside the building on 17 June 1972 (see). See also 3 September 1971 for similarities in the purported "Fielding office break-in," including personnel involved and the use of a holiday weekend, in that case the Labor Day weekend.]

On the same weekend as the purported Watergate "first break-in," L. Ron Hubbard goes absent from his usual duties and activities in the company of Green Beret Paul Preston. He's reported to have "moved ashore."

The crew of the Scientology Flagship Apollo are told that L. Ron Hubbard has "moved onshore." His "bodyguard" purportedly is Green Beret Paul Preston.

[NOTE: From this date until his reported death in 1986, L. Ron Hubbard never makes another public appearance. His whereabouts generally are unknown except to a few close people, who later claim that while with them he had been either "in hiding" or "on the run" or ill the entire time, including donning various disguises.] [85] [62]

Monday, 29 May 1972 (Memorial Day)

Ingo Swann is told by psychiatrist Karlis Osis that there are to be "no more remote viewing experiments at the ASPR." No reason is given. Swann calls fellow Scientology OT VII Hal Puthoff at SRI and offers to come out. [83] [77] [51] [44]

Sunday, 4 June 1972

OT VII Ingo Swann flies from New York to San Francisco, where he is met by OT VII Hal Puthoff and taken to SRI. [44]

Tuesday, 6 June 1972

Ingo Swann mentally affects a super cooled magnetometer encased in solid concrete five feet beneath the foundation of the Varian Hall of Physics, Stanford University, witnessed by Dr. Arthur Hebbard, Dr. Marshal Lee, and representatives of CIA. [44]

Wednesday, 7 June 1972

Willis Harmon meets OT VI Ingo Swann at SRI and takes Swann to a meeting where there are 16 people. Harmon is Director of his own Educational Policy Research Center at SRI, a center for "Futurology."

At the time, futurology constitutes one of the most important and biggest efforts in the world, and Harmon is well connected in Washington, D.C., with offices there. Harmon explains to Swann at the meeting that part of their ongoing project is to see if parapsychology and/or psychic abilities can or should be factored into "future scenarios." Harmon explains that all was known about the ASPR goings-on, and that the attempt to expel Swann "gives you more credentials than you realize, and also makes it easier for various people."[44]

Thursday, 8 June 1972

Ingo Swann goes to the home of Kirlian researcher Bill Tiller and there meets psychiatrist Shafica Karagulla.[44]

Friday, 9 June 1972

OT VII Ingo Swann leaves SRI and returns to New York City.

John Paul Vann—who had been closely involved with Lucien Conein and Daniel Ellsberg in Vietnam contemporaneously with Green Beret Paul Preston—is killed in a bizarre helicopter crash in Vietnam.

G. Gordon Liddy purportedly has a private meeting with Magruder where they purportedly discuss problems with "the room monitoring device" in the DNC and the prospects of "another entry" into the Watergate.

[NOTE: There is no "room monitoring device" in the DNC. See Watergate first break-in.] [44] [86] [53]

Monday, 12 June 1972

OT VII Ingo Swann agrees to return to the ASPR "for further research and experiments."[44]

Jeb Magruder purportedly has another private meeting with Liddy and orders Liddy to "go back into Watergate."[79] [53]

The telephone company sweep of Democratic National Committee headquarters in the Watergate finds no trace of bugs that Watergate burglars later will claim they had planted.

Thursday, 15 June 1972

The telephone company does a sweep of Democratic National Committee Headquarters in the Watergate. No electronic bugging devices are found. [NOTE: For full coverage, see Watergate first break-in.] [118]

Saturday, 17 June 1972

Five burglars are arrested at 2:30 a.m. in Democratic National Committee headquarters in the Watergate: James McCord, Bernard Barker, Frank Sturgis, Eugenio Martinez, and Virgilio Gonzalez. All five men have a history of being employed by CIA. CIA veteran James McCord has had to tape a door latch twice to get them arrested. They have bugging equipment with them, and several of the men have in their possession amazingly traceable $100 bills that will trace back to the White House. Bernard Barker has the phone number of E. Howard Hunt on him, indicating another connection to the White House.

CIA Watergate Goon Platoon: Hunt, McCord, Barker, Baldwin, Martinez, Sturgis, Gonzalez, and Liddy.

Director of CIA and convicted perjurer Richard Helms says: "We know the people...But there is no CIA involvement."

Almost at once the men start claiming to authorities that they had broken in weeks earlier, on 28 May 1972 (see), and were there to "fix" failures from the purported "first break-in," mainly electronic bugs.

[NOTE: But there was no "first break-in," and the phone company had just days before found there were no bugs in DNC headquarters. See Watergate first break-in. The amazing amount of obvious evidence on the men soon leads investigators to Liddy, Hunt, and Alfred Baldwin, who also are linked to the purported Memorial Day weekend "first break-in," providing them with an alibi for their whereabouts during that weekend.]

CIA Director Richard Helms claims to have been "preparing for bed" (at 3:00 a.m.?) when he gets a call from CIA Chief of Security Howard Osborn informing Helms that "District police" have picked up five men in a break-in. Helms is told that James McCord is one of them, along with "four Cubans." Osborn also purportedly tells Helms that "Howard Hunt also seems to be involved in some way." Helms purportedly asks Osborn: "Is there any indication that we could be involved in this?" and is told "None whatsoever." Next, "still sitting on the edge of the bed," Helms calls Acting Director of the FBI L. Patrick Gray, who is "in a Los Angeles hotel room." Gray says that he's been informed of the break-in, but has no details. Helms tells Gray that "despite the background of the apparent perpetrators, CIA had nothing to do with the break-in."[87]

Sunday, 18 June 1972

Ingo Swann flies to Northfield, Minnesota to give lectures at the annual retreat of Spiritual Frontiers Fellowship (SFF). [44]

Thursday, 22 June 1972

Charles Colson is interrogated by the FBI on the Watergate break-in. After interrogating Colson, the FBI is of the belief that the break-in is "a CIA thing."

Acting FBI Director L. Patrick Gray has a meeting at about five o'clock with SA Bates, the Assistant Director in charge of the General

Investigative Division of the FBI. Following the meeting, Gray places a telephone call to Richard Helms, Director of CIA, to "tell him our thought that we may be poking into a CIA operation in connection with the Watergate burglary." Helms tells Gray that Helms has been "meeting with his men on this every day," and that "although we know the people, we cannot figure this one out. But there is no CIA involvement." Helms then meets with Gray and "asks" Gray "not to interview the two CIA men." Gray issues the order. Gray calls FBI SA Bates "immediately following that visit" from Helms, and tells Bates that "there was some CIA involvement here," that "we should proceed very gingerly and very discreetly and carry out the investigation at the Banco Internationale, and also continue to try to trace these checks through the correspondent banks, but to hold off interviewing Mr. Ogarrio." Later that evening, Gray meets with John Dean. He tells Dean that Richard Helms has said "there is no CIA involvement."[88][22]

Friday, 23 June 1972

10:04 to 11:39 a.m.: In an Oval Office conversation, President Richard Nixon says "...the FBI agents who are working the case, at this point, feel that's what it is—this is CIA. ...[W]e protected Helms from one hell of a lot of things. ...This involves these Cubans, Hunt, and a lot of hanky-panky that we have nothing to do with ourselves." Ehrlichman answers that after interviewing Charles Colson the FBI "are now convinced it is a CIA thing."

11:06 a.m.: Acting FBI Director L. Patrick Gray has a phone conversation with John Dean. Dean tells Gray that if the FBI persists in investigating the Mexican money chain they will "be uncovering or become involved in CIA operations." Gray tells Dean that CIA Director Richard Helms told Gray the day before that "there is no CIA involvement" in the Watergate break-in. Gray also tells Dean, "if there is CIA involvement, let the CIA tell us."

[NOTE: Nixon and Haldeman are still in their meeting, which goes until 11:39 a.m.]

Acting FBI Director L. Patrick and the CIA waltz. Gray soon will destroy crucial evidence taken from the White House safe of CIA's golden boy, E. Howard Hunt.

2:19 p.m.: Dean calls Gray to find out if Gray has made an appointment with Deputy CIA Director Vernon Walters. Gray hasn't. Dean tells Gray that Walters will be calling Gray for an appointment and Gray should see him.

2:20 to 2:45 p.m.: Haldeman reports to Nixon that he and Ehrlichman [and John Dean] have met with CIA Director Helms and Deputy Director Vernon Walters. Helms has said, "We'll be very happy to be helpful," but Walters has said, "I don't know whether we can do it." Walters, though, is going to put in a call to Patrick Gray.

2:35 p.m. Vernon Walters meets with L. Patrick Gray. Walters says that if the FBI proceeds with the investigation into the Mexican money chain, they "would uncover CIA assets and resources" and could "interfere with some CIA covert activities." Walters then mentions to Gray "the agreement between the agencies not to uncover one another's sources," saying further that the FBI has "the five people and that the matter ought to be tapered off there."

2:53 p.m. After the meeting with Walters, Gray calls John Dean and tells Dean that Walters has indicated that there is "some CIA involvement," and that they will "proceed very gingerly and very discreetly and work around this until we can determine what we have a hold of."[89] [22] [88]

On the same day, an airgram is sent from the American Embassy in Copenhagen to the U.S. State Department from the Legal Attaché (LEGAT) Copenhagen titled "SUBJECT: L. RON HUBBARD FPC ." Its contents are unknown.

[NOTE: The only record of this airgram is a later memorandum, dated 5 September 1972 (see), to Acting Director of FBI L. Patrick Gray, enclosing a copy of the airgram, saying it is "self-explanatory."]

Saturday, 24 June 1972

According to Ingo Swann, he arrives in Washington, D.C. from Minnesota, ostensibly to "do book research at the Library of Congress"—but Swann says elsewhere that his trip to D.C. in 1972 was "to discuss psi phenomena with a variety of officials."[44]

Tuesday, 27 June 1972

Hal Puthoff contacts K. Green, Office of Strategic Intelligence (OSI) at CIA, informing Green of the results of the Varian Hall magnetometer experiment with Ingo Swann. There are also subsequent conversations between Puthoff and CIA personnel regarding this event. [19]

Wednesday, 28 June 1972

L. Patrick Gray gets a call from CIA Director Richard Helms, who asks Gray "not to interview active CIA men Karl Wagner and John Caswell." Gray immediately orders "that the interviews of John Caswell and Karl Wagner be held in abeyance." Caswell and Wagner's names have been found in a telephone-address notebook belonging to E. Howard Hunt.

In the evening, John Dean turns over some of the items from the White House safe of E. Howard Hunt to Gray. Gray is provided with a large brown envelope to carry the items away in. Dean tells Gray that included papers have "national security implications," saying they should "never see the light of day." Gray purportedly never looks at the papers, but takes them to his apartment in Washington D.C. and puts them on a closet shelf under his shirts.

Gray has a meeting with Mark Felt and SA Bates on "the CIA ramifications."[88]

Friday, 30 June 1972

Scientology OT VII Hal Puthoff says in a letter to Dr. Gertrude Schmeidler in New York that he has "obtained a contract to investigate the primary perception hypothesis of Cleve Backster."[44]

Saturday, 1 July 1972

The classified Defense Intelligence Agency (DIA) report entitled "Controlled Offensive Behavior—USSR" is published, though its findings have been known by top personnel for months. In part, it states: "The Soviet Union is well aware of the benefits and applications of parapsychology research. The term parapsychology denotes a multi-disciplinary field consisting of the sciences of bionics, biophysics, psychophysics, psychology, physiology and neuropsychology.

Celebrity Centre's Yvonne (Gillham) Jentzsch has standing orders to be located for any incoming calls from Puthoff or Swann

Many scientists, U.S. and Soviet, feel that parapsychology can be harnessed to create conditions where one can alter or manipulate the minds of others. The major impetus behind the Soviet drive to harness the possible capabilities of telepathic communication, telekinetic and bionics are said to come from the Soviet military and the KGB [Committee of State Security; Secret Police]. ...Soviet knowledge in this field is superior to that of the U.S. ...The potential applications of focusing mental influences on an enemy through hypnotic telepathy have surely occurred to the Soviets.... Control and manipulation of the human consciousness must be considered a primary goal. ...Soviet efforts in the field of psi research, sooner or later, might enable them to do some of the following: (a) Know the contents of top secret US documents, the movements of our troops and ships and the location and nature of our military installations (b) Mold the thoughts of key US military and civilian leaders at a distance (c) Cause the instant death of any US official at a distance (d) Disable, at a distance, US military equipment of all types, including spacecraft."[71]

Yvonne Gillham Jentzsch, Executive Director of Scientology's Celebrity Centre, is married to Heber Jentzsch. Despite running an organization with over 200 staff members and a grueling schedule, including appearances around the U.S. and several foreign countries, she has standing orders with her office and public relations staff to locate her wherever she is if a call should come in for her from OT VIIs Hal Puthoff or Ingo Swann.

[NOTE: According to staff members who contributed the information, Yvonne Jentzsch had no specific knowledge at the time of Swann or Puthoff's connections with CIA or NSA, only that they both had contact with various influential people, and possibly even was of the belief that their connections were related somehow to NASA and the space race, but not to military intelligence.]

Monday, 3 July 1972

According to one of several conflicting accounts told by L. Patrick Gray, he burns the papers given to him by John Dean that had been taken from the safe of E. Howard Hunt in a wastebasket in his office at the FBI.

[NOTE: Gray later retracts this story, saying that he kept the papers first in his apartment, then moved them to his office, then to his home, where he burned them on or around 27 December 1972 (see).]

Gray has another meeting with Mark Felt, Bates, and also "Mr. Kunkel, the Special Agent in charge of the Washington Field Office" on "the CIA ramifications."[88]

Monday, 17 July 1972

Over $1.1 million in cash never accounted for except by ledger "credit" years later, during IRS's Meade Emory restructuring of Scientology

A sum equivalent to US $1,119,678 in Swiss francs is withdrawn in cash by Fred Hare and Vicki Polimeni from a trust fund (of questionable origin) in Switzerland and purportedly is brought aboard the Flagship Apollo and put into a safe.

[NOTE: Conflicting accounts in the same referenced Tax Court ruling say that the amount was "over $2 million," and also say the cash was put into "a file cabinet in a strongroom" instead of a safe. The same ruling also provides no accounting of what happened to the actual cash.] [90]

Wednesday, 19 July 1972

Fred LaRue gives $40,000 to Herbert W. Kalmbach, who takes it to New York and gives it to Anthony Ulasewicz.

Ulasewicz delivers $40,000 to Dorothy Hunt—wife of E. Howard Hunt—and $8,000 to G. Gordon Liddy in unmarked envelopes left in lockers at Washington National Airport. [91]

Cash to Dorothy Hunt, wife of CIA's E. Howard Hunt, and more cash to G. Gordon Liddy

Wednesday, 26 July 1972

A report is issued entitled "Report of an Out-of-Body Experiment Conducted at the American Society for Psychical Research: Participants: Dr. Carole Silfen, Janet Mitchell, Ingo Swann." The report describes an OOB experiment that suggests that a point of perception exterior to the body is able to assume "at a different location the functions performed by the visual system and the brain in the body." This is the first such experiment that verified the capability of such remote points of view.[44]

Monday, 7 August 1972

Unknown amount of cash delivered by NSA's Hal Puthoff to Ingo Swann

OT VII Ingo Swann flies to San Francisco and is met by OT VII Hal Puthoff. Puthoff gives Swann an envelope containing an unspecified amount of cash, and a copy of their three-week schedule. They are to have a one-week informal period, and then a two-week formal set-up. The latter two-week segment will be attended by two CIA representatives. [44]

Friday, 11 August 1972

Ingo Swann flies to Los Angeles for the weekend with psychiatrist Shafica Karagulla and "her associate," even though he has come to SRI specifically to perform experiments in the presence of CIA personnel. No reason is given for the trip.

[NOTE: Karagulla is a neurosurgeon. and has studied under Canadian psychiatrist Wilder Penfield, infamously known for putting electrical probes into the brains of conscious subjects.] [44]

Monday, 14 August 1972

Ingo Swann's weekend travelling companion Shafica Karagulla, a psychiatrist and neurosurgeon, shown here with her mentor, psychiatrist Wilder Penfield

Swann is back at SRI, after his trip to Los Angeles with psychiatrist Shafica Karagulla, and is ready to begin the two-week formal experiments in the company of two representatives from CIA. [44]

Wednesday, 23 August 1972

A CIA project officer contracts Hal Puthoff for a demonstration with OT VII Ingo Swann. Swann is asked to describe objects hidden out of sight by CIA personnel. The descriptions are so "startlingly accurate" that Swann purportedly is asked if he will complete the necessary forms "for a security clearance."

[NOTE: Swann is already on record as having a top secret clearance.]

He agrees to do it once he gets back to New York "where his papers are." The CIA rep suggests to CIA that the work be continued and expanded. CIA's Sidney Gottlieb reviews the data, approves another work order, and encourages the development of "a more complete research plan."[19]

Saturday, 26 August 1972

Ingo Swann returns to New York from SRI. He prepares the application for security clearance and sends it off to Hal Puthoff. [44]

Wednesday, 30 August 1972

A once-sentence letter is received by the FBI. It says: "Did you receive the printed matter that was sent to you concerning Scientology, if so please acknowledge. Thank you."

[NOTE: In the released FBI copy, the signature is blacked out. The letter is answered two days later (see 1 September 1972) by Acting FBI Director L. Patrick Gray.]

September 1972

The Scientology Flagship Apollo is moved to Spain for refit. The crew and officers are given the story that L. Ron Hubbard is still "living ashore" to account for his absence.

L. Ron Hubbard has functionally disappeared, his purported whereabouts known only to a small number of people called the "Special Unit" (SU).

Friday, 1 September 1972

Acting Director of FBI L. Patrick Gray responds to a once-sentence letter received by the FBI received two days earlier (see Wednesday, 30 August 1972). Gray's reply says: "Your letter was received on August 30th. With respect to your inquiry, a search of our records does not reveal any prior communication from you."

[NOTE: In the released FBI copy, the address block and the person's name is blacked out, and the letter has a note at the bottom: "Correspondent is not identifiable in Bufiles."]

Tuesday, 5 September 1972

Acting Director of FBI L. Patrick Gray receives a memorandum from the Legal Attaché (LEGAT) Copenhagen (163-222) (RUC) titled "SUBJECT: L. RON HUBBARD FPC regarding an airgram sent to the State Department on 23 June 1972 (see). It says: "ReCOPlet 5/23/72. Enclosed are single copies [sic] of an airgram dated 6/23/72, captioned "THE CHURCH OF SCIENTOLOGY IN DENMARK," from AmEmbassy, Copenhagen, to U.S. Dept. of State, which is self-explanatory."

[NOTE: Copies go to Foreign Liaison, Legat Madrid, and Copenhagen]

Friday, 15 September 1972

Hunt, Liddy, McCord and the Watergate burglars are indicted by a federal grand jury. The involvement of McCord and Liddy provide investigators with a link to the Nixon campaign. The involvement of E. Howard Hunt provides investigators with a link to the White House. [92]

Tuesday, 19 September 1972

More cash to Dorothy Hunt, wife of CIA's E. Howard Hunt

Anthony Ulasewicz flies to Washington, D.C. and delivers $53,000 cash to Dorothy Hunt—wife of E. Howard Hunt—and $29,000 to Fred LaRue by leaving unmarked envelopes in a locker at Washington International Airport and in the lobby of a motel near LaRue's residence. [91] [93]

Saturday, 30 September 1972

According to one of the conflicting stories he told, at the end of September Acting Director of the FBI L. Patrick Gray takes files that had been in E. Howard Hunt's White House safe to his home in Stonington, Connecticut, and puts them in a chest-of-drawers intending to burn them. [88]

Sunday, 1 October 1972

On a Sunday, CIA's Technical Services Division (TSD) awards OT VII Hal Puthoff a top-secret research contract to develop "remote viewing" for military espionage purposes.

[NOTE: TSD is the CIA division formerly known as "Technical Services Staff." TSD is also the division running MK-ULTRA. The head of TSD is Sidney Gottlieb. The name of TSD will change a month after this contract to "Office of Technical Services." Its acronym, OTS is a pun.] [19][94]

November 1972

CIA Director Richard Helms calls L. Patrick Gray's "number two man," Mark Felt, stating that Helms is going to call Assistant Attorney General Peterson regarding the interview of CIA's Karl Wagner to see if it "could not be conducted...be held off."

CIA's Sidney Gottlieb "retires."

The name of CIA's TSD is changed to Office of Technical Services (OTS). [88] [19]

OT III Pat Price

Sunday 3 December 1972

L. Ron Hubbard purportedly "goes into hiding" in New York in the company of Green Beret Paul Preston.

Scientology OT VIIs Hal Puthoff and Ingo Swann, now under contract with CIA, "run into" Scientology OT III Pat Price, who purportedly is selling Christmas trees at a lot in Mountain View, California—close to SRI. Puthoff is reported to "have met" Price "several years earlier" at a lecture in Los Angeles.

[NOTE: Los Angeles is the location of Scientology's Advance Organization Los Angeles (AOLA), the only place in the U.S. at the time where the OT Levels are delivered.][94]

Friday, 8 December 1972

E. Howard Hunt's wife Dorothy is killed in a plane crash in Chicago

E. Howard Hunt's wife, Dorothy Hunt, is killed in the United Airlines airplane crash of Flight 533 as it approaches Chicago. Dorothy Hunt's purse contains $10,585 cash, most of it in hundred dollar bills.

Thursday, 21 December 1972

OT VII Ingo Swann arrives to begin his CIA contract at Stanford Research Institute.

James W. McCord writes a letter to Jack Caulfield that says in part: "If Helms goes, and if the WG (Watergate) operation is laid at the CIA's feet, where it does not belong, every tree in the forest will fall. It will be a scorched desert. The whole matter is at the precipice right now. Just pass the message that if they want it to blow, they are on exactly the right course." Caulfield replies: "I have worked with these people and I know them to be as tough-minded as you. Don't underestimate them."[94] [44][95]

1973

January 1973

CIA Director Richard Helms orders that records of CIA's OTS, including records relating to MK-ULTRA, be deliberately destroyed. [96]

Wednesday, 3 January 1973

On the very day that Ellsberg goes on trial, the CIA hand-couriers to Watergate prosecutors CIA's own copies of photos of Liddy and Hunt in front of Fielding's office

Daniel Ellsberg goes on trial, accused of theft and conspiracy in the disclosure of the Pentagon Papers.[97]

On the same day, CIA's Anthony Goldin hand delivers to the Department of Justice Watergate Prosecutors copies of 10 photos of E. Howard Hunt and G. Gordon Liddy taken at the office of Ellsberg psychiatrist Lewis J. Fielding, with Fielding's name on the door clearly visible.

[NOTE: See 26 August 1971, when Liddy and Hunt flew to Los Angeles to take the photos of each other.] [93]

Thursday, 4 January 1973

Jack Caulfield delivers to John Dean a handwritten copy of the 21 December 1972 letter Caulfield had received from James McCord: "If Helms goes, and if the WG (Watergate) operation is laid at the CIA's feet, where it does not belong, every tree in the forest will fall. It will be a scorched desert."[93]

Thursday, 1 February 1973

A translation of a Soviet paper, "Report from Movosibirsk: Communication between Cells," appears in Vol. 7, No. 2 of the Journal of Parapsychology. The report says that experiments conducted in "Special Department No. 8" indicate that cells could communicate illness, such as a virus infection, despite the fact the cells are physically separated. The tests showed that when one group of cells was contaminated with a virus, the adjacent group—although separated by quartz glass—"caught the disease."[17]

Thursday, 1 February 1973

OT VII Hal Puthoff and Russell Targ have meetings with "selected Agency [CIA] personnel" at CIA headquarters in Langley, Virginia, to review the results of their research contract with CIA. Several Office of Research and Development officers show interest in contributing their own "expertise and office funding" to the research efforts. Prior to this, the contract has been with CIA's Office of Technical Services (OTS).[19]

Wednesday, 7 February 1973

CIA Director Richard Helms perjures himself before the Senate Foreign Relations Committee about CIA attempts to overthrow the government in Chile. [98]

Mid February 1973

CIA's Office of Research and Discovery (ORD) sends Project Officers to SRI to report on the remote viewing experiments of OT VIIs Hal Puthoff and Ingo Swann. ORD is considering joining CIA's Office of Technical Services in sponsoring the research on a joint program basis. [19]

Tuesday, 6 March 1973

Senator Frank Church will later head 1975 investigations into U.S. intelligence agency crimes—but he never exposes CIA's remote viewing program or its genesis.

Richard Helms testifies before the Multinational Committee, headed by Frank Church. Nothing about CIA's remote viewing program is revealed. [98]

Saturday, 17 March 1973

John Dean informs President Nixon that the Watergate Committee has learned that Justice Department prosecutors have CIA-supplied photos of E. Howard Hunt and G. Gordon Liddy taken at the office of Ellsberg psychiatrist Lewis J. Fielding. It's the first Nixon has heard anything about the Hunt/Liddy operation or of the photos. Dean says of Hunt and Liddy: "These fellows had to be some idiots."

[NOTE: See 26 August 1971 and 3 January 1973][64]

On or about the same day, E. Howard Hunt meets with Paul O'Brien, an attorney for the Committee to Re-elect the President. He tells O'Brien that "commitments had not been met," that he has done "seamy things for the White House," and that unless he receives $130,000, he "might review his options."[93]

Wednesday, 21 March 1973

$75,000 cash is delivered to CIA's E. Howard Hunt

Fred LaRue arranges for $75,000 in cash to be delivered to E. Howard Hunt through Hunt's attorney, William Bittman.

April 1973

According to CIA Project Officer over the SRI experiments, Ken Kress, it's about this time that Scientology OT III Pat Price starts working with OT VIIs Puthoff and Swann at SRI, and that "the remote viewing experiments in which a subject describes his impressions of remote objects or locations began in earnest."

[NOTE: Hal Puthoff asserts that Pat Price came on board at SRI 1 June 1973.][19]

Sunday, 15 April 1973

Assistant Attorney General Henry Petersen supplies to Judge Matthew Byrne—the judge in the Daniel Ellsberg Pentagon Papers trial—copies of the CIA-supplied photos of E. Howard Hunt and G. Gordon Liddy in front of the office of Ellsberg's psychiatrist, Lewis Fielding. [93]

John Dean spills the beans on the "break-in" at psychiatrist Lewis J. Fielding's office. All charges soon will be dropped against Ellsberg.

On the same day, John Dean tells federal prosecutors about the burglary of Dr. Lewis Fielding's office in Los Angeles, engineered by E. Howard Hunt.[98]

OT VII Ingo Swann comes up with the name "coordinate remote viewing" instead of just "remote viewing."[94]

Monday, 16 April 1973

E. Howard Hunt confirms what John Dean has told federal prosecutors the previous day about the burglary of psychiatrist Lewis Fielding's office in Los Angeles.[98]

Friday, 20 April 1973

CIA's Office of Research and Development decides to become involved in the remote viewing research, requests an increase in the scope of the effort, and transfers funds to CIA's Office of Technical Services (OTS): "C/TSD; Memorandum for Assistant Deputy Director for Operations; Subject: Request for Approval of Contract; 20 April 1973 (SECRET)."[19]

Late April 1973

OT VII Ingo Swann's "coordinate remote viewing" experiments are getting more accurate and promising results, prompting Hal Puthoff and Russell Targ to continue the experiments.[94]

Early May 1973

CIA Project Officer Ken Kress is told not to increase the scope of the SRI remote viewing research because it's "too sensitive": CIA's Office of Technical Services (OTS) is being investigated for involvement in the Watergate affair.[19]

Director of Central Intelligence Dr. James Schlesinger issues a memorandum to all CIA employees requesting the reporting of any activities that may have been illegal and improper: CIA operations are being investigated in connection with Watergate.[19]

Friday, 11 May 1973

Because of CIA-supplied photos of G. Gordon Liddy and E. Howard Hunt at the office of Daniel Ellsberg psychiatrist Lewis Fielding and the subsequent revelations, all charges against Daniel Ellsberg for leaking the Pentagon Papers are dropped and his case dismissed on the grounds of "government misconduct."[99]

Daniel Ellsberg is cleared of all charges because of CIA-supplied photos—that had been taken with a CIA camera and developed by CIA—of Liddy and Hunt in CIA garb in front of psychiatrist Lewis Fielding's office.

Monday, 21 May 1973

A twenty-six-page preliminary summary of the reports from CIA employees regarding questionable activities is sent to DCI William Colby under the title "Potential Flap Activities." The full report, completed later, comes to 693 pages in all, one for each "abuse," and it quickly acquires the nom de scandale of "the Family Jewels." It includes reports on CIA's involvement in assassination plots.[98]

Tuesday, 29 May 1973

CIA analyst Richard Kennet gives a set of coordinates he's gotten from a CIA colleague to Hal Puthoff at SRI as a "rigorous scientific experiment," coordinates that Kennet himself doesn't know anything about. [94]

Early June 1973

OT VII Ingo Swann and OT III Pat Price do remote viewing sessions on the coordinates given to Hal Puthoff by CIA's Richard Kennet. Their results are similar, both sketching something that resembles some sort of military installation. Price's report is detailed, including even code names on file folders on desks and inside file cabinets, and names of military personnel. Puthoff sends the information off to CIA's Richard Kennet. [94]

Friday, 8 June 1973

Richard Kennett shows Pat Price's and Ingo Swann's "coordinate remote viewing" results to his CIA colleague, Bill O'Donnell, who had provided Kennett the coordinates in question to begin with [see timeline entry for 29 May 1973]. O'Donnell it isn't even close; he had given Kennett map coordinates for his summer cabin in the woods. [94]

Sunday, 10 June 1973

Richard Kennett takes his wife and children on a "drive into the countryside" in the Blue Ridge Mountains to check out Bill O'Donnell's accounting of the coordinates Pat Price and Ingo Swann had remotely viewed. "A few miles from his friend's cabin," Kennett discovers a dirt road with a government "No Trespassing" sign, and some satellite antennas in the background—"obviously some kind of secret installation." It seems to match many of the descriptions provided by Price and Swann. [94]

Monday, 11 June 1973

Richard Kennett looks up "an official who he thought might know about" the strange secret base he and his wife and kids have discovered on their weekend drive to West Virginia, and gives the unnamed official Pat Price's and Ingo Swann's descriptions from their "coordinate remote viewing" sessions. [94]

Wednesday, 13 June 1973

CIA's Richard Kennett finds himself at the center of an intense and hostile security investigation over the "coordinate remote viewing" descriptions of Scientology OTs Pat Price and Ingo Swann of the secret installation in West Virginia. The investigation soon extends to Price, Swann, and OT VII Hal Puthoff at SRI. The facility, ostensibly a U.S. Navy communications base, is actually a highly sensitive NSA installation. [94]

The NSA's David Young—who has been granted immunity by Watergate prosecutors—turns over to them a one-page memo revealing a 1971 plan for G. Gordon Liddy and E. Howard Hunt to arrange a break-in at the office of Ellsberg psychiatrist Lewis J. Fielding. [100]

Late June 1973

Hal Puthoff and Russell Targ brief senior CIA officials at CIA Headquarters in Langley, Virginia on their remote viewing research. The officials include Office of Technical Services (OTS) chief John McMahon and Deputy Director for Science and Technology Carl Duckett. [94]

July 1973

Hal Puthoff and Ingo Swann travel to Prague for the First International Congress on Psychotronic Research. Word comes from CIA that the leader of the Soviet group is a KGB officer. At the same conference is CIA's Cleve Backster. [94] [101]

August 1973

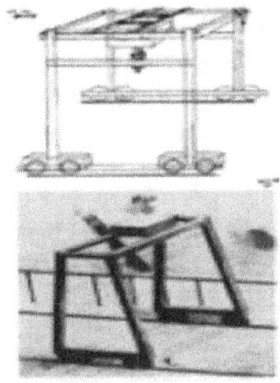

TOP: OT III Pat Price remote viewing sketch of Soviet gantry at Semipalatinsk.

BOTTOM: CIA illustration of gantry made from satellite photos.

OT III Pat Price is given coordinates supplied by CIA's Ken Kress for coordinate remote viewing experiments. Price identifies a Soviet military research facility at the southern edge of the Semipalatinsk nuclear test area in the Kazakh Republic. The accuracy of Price's reports about the place become an important factor in future funding of the remote viewing research of Puthoff, Targ, et. al. [94]

CIA officials discuss parapsychology with "several members" of DIA (Defense Intelligence Agency). The "DIA people" are interested in the Soviet activities in this area, and express considerable interest in the CIA/SRI experiments. [19]

Ingo Swann's contract at SRI ends. He returns to New York.

[NOTE: Later timeline entries indicate that Ingo Swann has been training CIA "in-house" remote viewers.] [94]

October 1973

Ingo Swann is flying from New York to Los Angeles every weekend to receive Scientology services at Celebrity Centre from Jim Fiducia. Swann has completed a service called "Grade IV Expanded." He writes a Scientology "success story" that says in part: "The precision of Ron's (L. Ron Hubbard's) auditing technology...is such a great contribution to history and humanity that words are not enough. Utilizing the technology is what to do 'in this point in time.'"[102]

Early November 1973

William Colby has become Director of Central Intelligence (DCI). [19]

Ingo Swann, developing remote viewing for CIA, says: "The precision of [L. Ron Hubbard's] technology is such a great contribution to history and humanity that...utilizing the technology is what to do in this point in time."

Friday, 9 November 1973

CIA's K. Green issues a report on the 1 June 1973 [see] coordinate remote viewing experiment with Ingo Swann and Pat Price that had targeted a secret NSA installation in West Virginia: "K. Green; LSD/OSI; Memorandum for the Record; Subject: Verification of Remote Viewing Experiments at Stanford Research Institute; 9 November 1973. (SECRET)." The "new directors" of CIA's Office of Technical Services and Office of Research and Development are favorably impressed. [19]

Late November 1973

Based on the favorable impression made by the 9 November 1973 "Verification of Remote Viewing" report, a CIA Statement of Work is outlined, and the SRI team (Puthoff, Targ, et al.) is asked to propose another program. [19]

1974

January 1974

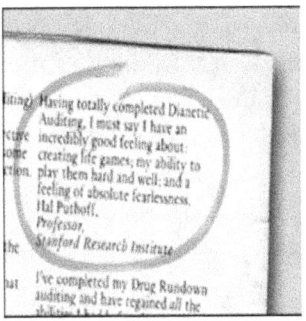

NSA's Hal Puthoff doing Scientology services at Celebrity Centre as a "professor." (Scans contributed, from Celebrity magazine Minor Issue 9, circa February 1974.)

NSA's Hal Puthoff, a Scientology OT VII contracted to CIA, has completed "Dianetic Auditing" at Scientology's Celebrity Centre in Los Angeles. His "Success Story" soon appears in Celebrity magazine saying he has "a feeling of absolute fearlessness." He has represented himself to the Scientologists as merely a "Professor, Stanford Research Institute."

Somehow the Department of Justice and FBI have upper level, confidential Scientology (OT) materials in their files. [103]

Friday, 1 February 1974

A new CIA program, jointly funded by Office of Research and Development (ORD) and Office of Technical Services (OTS) is begun at SRI. Kenneth Kress is the author Project Officer of the program. The project is to proceed on the premise that the phenomena associated with remote viewing exist; the objective is to develop and utilize them. The program is referenced by a cite: "Office of Technical Services Contract, FAN 4125-4099; Office of Research and Development Contract, FAN 4162-8103; 1 February 1974 (CONFIDENTIAL)." [19]

Tuesday, 5 February 1974

Hal Puthoff is contacted by the Berkeley police, requesting psychic assistance in the investigation into the disappearance of Patty Hearst. That afternoon, Puthoff and Pat Price meet with the police at Patty Hearst's apartment, where Price says it is not a kidnapping for money, but for political reasons.

Pat Price, a Scientology OT III involved in secret CIA remote viewing program, is called in to help on the Patty Hearst kidnapping

They go down to the police station, where Price picks out three photos from mug books, and associates the word "Lobo" (spanish for "wolf") with one of the men he has selected. (All three men Price picked are later confirmed to be members of the "Symbionese Liberation Army," which has kidnapped Hearst. The man with the "lobo" association turns out to be William Wolfe, a.k.a. "Willie the Wolf.") [94]

Saturday, 17 August 1974

Ingo Swann gives a lecture to about 250 Scientologists at Laurel Springs Ranch in Santa Barbara, California: "What has Scientology got to do with Psychic Research?" Its topics include, "What is a Spirit? Its Potentials, and How Scientology provides a workable way for anyone to know the answers for himself." Swann has flown in from New York for the lecture—where Swann secretly has been training CIA personnel in Scientology-based remote viewing. [104]

Tuesday, 20 August 1974

A secret internal CIA report is issued regarding OT III Pat Price's August 1973 [see] remote viewing of a Soviet Research and Development facility: "W. T. Strand; C/ESO/IAS; Memorandum for Director, Officer of Technical Service; Subject: Evaluation of Data on Semipalatinsk Unidentified R&D Facility No. 3, USSR; 20 August 1974 (SECRET)." [19]

Thursday, 29 August 1974

Scientology legal has filed a Freedom of Information Act (FOIA) request to the U.S. Treasury Department that becomes civil case No. 76-1719, CSV v. Secretary of the Treasury, in U.S. District Court for the District of Columbia.

[NOTE: The case will be considerably expanded, entailing hundreds of documents, and including the Secret Service. Although some documents are ultimately released to the Scientologists, many are withheld under "the penumbra of agency's executive privilege which exempts from FOIA the decision-making processes of government agencies" and under protected "inter-agency or intra-agency memorandums or letters."]

Early October 1974

Ingo Swann returns to SRI as a "consultant."

[NOTE: Later timeline entries indicate that Ingo Swann has been training CIA "in-house" remote viewers.] [94]

Friday, 18 October 1974

A paper by OT VII Hal Puthoff and Russell Targ appears in Nature magazine: "Information Transfer Under Conditions of Sensory Shielding."[94]

Thursday, 12 December 1974

A CIA report: "CI/Staff/DDO; Memorandum for the Record; Subject: SRI Experiment; 12 December 1974 (SECRET)." [19]

Monday, 16 December 1974

NSA initially lies and says they have no documents on Scientology or Hubbard. NSA's Hal Puthoff, a Scientology OT VII, is running the secret remote viewing program for CIA.

The Founding Church of Scientology, Washington D.C. (FCDC) seeks access through FOIA to all records maintained by the National Security Agency (NSA) on FCDC and Scientology, as well as any records reflecting dissemination of information to other domestic agencies or foreign governments.

[NOTE: The action is soon expanded to include all references to other specific Scientology organizations and to L. Ron Hubbard. NSA claims in response that it has no records related to Scientology or Hubbard. That will turn out to be a lie, but the documents ultimately will be withheld on grounds of "national security" and "confidentiality specifically imparted by other statutes."]

Thursday, 19 December 1974

The Church of Scientology of California (CSC) files FOIA requests for 145 documents from the U.S. Treasury, the U.S. Secret Service, the Secretary of the Treasury, and the Director of Secret Service pertaining to Scientology and Scientologists and to L. Ron Hubbard.

Sunday, 22 December 1974

The New York Times publishes an article by Seymour Hersh regarding the secret Operation Chaos. Gerald Ford is President. DCI William Colby phones Ford, who is vacationing in Vail, Colorado, and tells him that Hersh has distorted the record, and that the "excesses" of CIA had all ended in 1973—following Helms's departure. [98]

1975

January 1975

Secret internal CIA reports are issued:

"AC/SE/DDO; Memorandum for C/D&E; Subject: Perceptual Augmentation Testing; 14 January 1975 (SECRET)"

"Chief/Division D/DDO; Memorandum for C/D&E; Subject: Perceptual Augmentation Techniques; 24 January 1975

"J. A. Ball; "An Overview of Extrasensory Perception"; Report to CIA, 27 January 1975.

"C/Libya/EL/NE/DDO; Memorandum for OTS/CB; Subject: Libyan Desk Requirement for Psychic Experiments Relating to Libya; 31 January 1975 (SECRET)"

"C/EA/DDO; Memorandum for Director of Technical Service; Subject: Exploration of Operational Potential of 'Paranormals'; 5 February 1975 (SECRET)" [19]

An internal CIA report is issued regarding the results of remote viewing experiments performed by CIA "insiders"—all members of CIA's Office of Technical Services (OTS): "OTS/SDB; Notes on Interviews with F. P., E. L., C. J., K. G., and V. C., January 1975 (SECRET)."

[NOTE: This is the first confirmation that CIA has their own in-house personnel as remote viewers. Later timeline entries indicate that Ingo Swann has been training CIA in-house remote viewers.] [19]

Meade Emory becomes Assistant to Commissioner of IRS. By 1982, Emory will restructure all of Scientology, putting it in the control of three lawyers who are not Scientologists

Around this time, Ingo Swann purportedly leaves Scientology: "I exited Scientology of my free will in 1975 and under reasonably amicable circumstances."

[NOTE: Unfortunately, Swann's claim is simply a lie. In August 1977 (see) he is one of the speakers listed for Scientology's "International Prayer Day," and in April 1979 (see) he is listed in a Scientology publication as having completed a service called "New Era Dianetics for OTs".]

Around the same time, FCDC expands its FOIA action against NSA to include all references to L. Ron Hubbard, founder of Scientology.

Around the same time, Meade Emory is appointed as Assistant to the Commissioner of IRS, Donald C. Alexander.

[NOTE: During Emory's tenure, a clerk typist named Gerald Wolfe will be hired at IRS despite a hiring freeze. Wolfe will begin feeding stolen documents to Scientology's Guardian Office, which later will be raided by FBI. The Guardian Office principals, including Mary Sue Hubbard, will be sent to jail. In the aftermath, Meade Emory engineers L. Ron Hubbard's probate, restructures all of Scientology, and becomes one of the Founders of Church of Spiritual Technology, the ultimate beneficiary of L. Ron Hubbard's estate.]

Friday, 14 February 1975

The NSA replies to FCDC's FOIA action that it has not established any file pertaining either to FCDC or L. Ron Hubbard, and that it has transmitted no information regarding either to any domestic agencies or foreign governments. [This proves later to be a bald-faced NSA lie.]

Thursday, 27 February 1975

CIA's William Colby on CIA assassinations: "Not in this country."

CBS correspondent Daniel Schorr, in a meeting with CIA Director William Colby, asks Colby point blank, "Has the CIA ever killed anybody in this country?" Colby responds: "Not in this country." Schorr is stunned at Colby's oblique admission, but Colby will not answer further questions about it, saying only that assassinations had been "formally prohibited in 1973." [NOTE: See 1972.] [98]

Early-mid March 1975

FCDC expands its FOIA action against NSA, naming other Scientology organizations that NSA is suspected of having documents on. NSA again denies possession of any of the data sought.

[NOTE: This, too, proves later to be a lie. As has been thoroughly covered, OT VII Hal Puthoff, running the secret CIA remote viewing program, is NSA.]

Around the same time, all CIA funding of remote viewing and paranormal research purportedly comes to a sudden halt. "To achieve better security," the "operations-oriented testing" of remote viewing with Scientology OTs Hal Puthoff and Ingo Swann purportedly is stopped. [19]

Around the same time, CIA "personal services" contract with Scientology OT III Pat Price is started. [19]

Pat Price departs SRI. He claims he is going to "work for a coal company" in Huntington, West Virginia, and intends to return in a year. He is working directly for CIA as a contractor. [94]

Thursday, 12 June 1975

Two internal CIA reports are issued regarding a device being used at SRI in remote viewing research:

"L. W. Rook; LSR/ORD; Memorandum for OTS/CB; Subject: Evidence for Non-Randomness of Four-State Electronic Random Stimulus Generator; 12 June 1975 (CONFIDENTIAL)."

"S. L. Cianci; LSR/ORD; Memorandum for OTS/CB; Subject: Response to Requested Critique, SRI Random Stimulus Generator Results; 12 June 1975 (CONFIDENTIAL)."[19]

Monday, 23 June 1975

Scientology FOIA actions against CIA expose NSA's lie about having no relevant documents. Both FOIA actions are ultimately thrown out by courts on grounds of "national security."

In the course of FOIA proceedings against the Department of State and the Central Intelligence Agency (CIA), FCDC learns that NSA has at least sixteen documents concerning Scientology, FCDC and related organizations despite NSA's claims for months that they had no such documents. Suddenly, confronted with details extracted by FCDC from the CIA, the NSA "succeeds" in locating fifteen of those items "in warehouse storage," and obtains a copy of the sixteenth from CIA. Then NSA takes legal action to prevent release of the materials on grounds of national security.

Friday, 11 July 1975

OT III Pat Price, on a "personal services contract" with CIA, is given a "second requirements list" for a Libyan installation Price had earlier identified with remote viewing as a guerilla training site. Price dies "a few days later." [19]

Tuesday, 15 July 1975

OT III Pat Price leaves from Huntington, West Virginia on a several-week trip west. He first stops and has dinner in Washington, D.C. [94]

Wednesday, 16 July 1975

OT III Pat Price arrives in Las Vegas, en route first to SRI, then to Los Angeles. In Vegas, Price is met by an old friend named Bill Alvarez and his wife, Judy. The three check into the Stardust Hotel and go into the restaurant for dinner. Price begins to complain that he doesn't feel well, and tells the Alvarezes that someone "had seemed to slip something into his coffee" at dinner in Washington the night before. Price soon feels so bad that he goes up to his room to lie down. He feels even worse and calls the Alvarezes. They come to his room and find him on the bed apparently in cardiac arrest. Bill Alvarez calls paramedics, who try without success to resuscitate Price. He is declared dead in the local hospital's emergency room. A mysterious "friend" of Price's turns up at the emergency room with "a briefcase full of his medical records," which, along with the statements of the emergency room's physician, are enough to waive an autopsy—which would normally be performed on an out-of-towner who had died outside the hospital. [94]

Tuesday, 9 September 1975

Although relevant records are still classified, there can be little doubt that the remote viewing program is going directly to benefit the Army, since the entire purpose was for military intelligence. This case, too, will be thrown out on grounds of "national security."

Pursuant to FCDC's FOIA requests, Department of Defense and Department of the Army have released a number of documents in their entirety, released only edited versions of others, and refused to release any portion of certain documents. Dissatisfied, the Church resorts to legal action to compel disclosure. On September 9, 1975, the Church files a complaint seeking an injunction against withholding of records: Church of Scientology v. United States Department of the Army, No. CV-75-3056-F. Named as co-defendants in the action are Secretary of the Army, the U. S. Intelligence Agency and Assistant Chief of Staff for Army Intelligence.

[NOTE: This FOIA case will be lost mainly on grounds of "national security."]

Wednesday, 8 October 1975

A CIA report is done regarding experiments being done at SRI: "G. Burow; OJCS/AD/BD; Memorandum for Dr. Kress; Subject: Analysis of the Subject-Machine Relationship; 8 October 1975 (CONFIDENTIAL)."[19]

A CIA report is issued that's somehow related to the "requirements list" for a Libyan remote viewing target that was allegedly passed to Pat Price just days before he died: "DDO/NE; Memorandum for OTS/BAB; Subject: Experimental Collection Activity Relating to Libya; 8 October 1975 (SECRET)." [19]

Thursday, 4 December 1975

The Defense Department also will be protected by U.S. Courts from releasing the documents on the grounds of "national security."

In addition to it 9 September FOIA suit, Scientology's FCDC files a complaint seeking an injunction against withholding of records in Church of Scientology v. United States Department of Defense, No. CV-75-4072-F. Named as co-defendants in the action are Office of the Secretary of Defense, Secretary of the Department of Defense, United States Department of the Navy, Secretary of the Navy, Naval Intelligence Command, and Director of Naval Intelligence.

[NOTE: This FOIA case, too, will be lost mainly on grounds of "national security."]

On the same day, an internal CIA report is issued on a Pat Price remote viewing of a Soviet Research and Development facility is issued: "D. Stillman; Los Alamos Scientific Laboratory; "An Analysis of a Remote

Viewing Experiment of URDF-3"; 4 December 1975 (CONFIDENTIAL)." [19]

1976

Wednesday, 14 January 1976

The AiResearch Manufacturing Company completes a report to CIA indicating that further developments in long-distance telepathy are continuing in the Soviet Union. [17]

Friday, 30 January 1976

George H.W. Bush becomes Director of CIA.

March 1976

OT VII Hal Puthoff and Russell Targ publish an article: "A Perceptual Channel for Information Transfer Over Kilometer Distances; Historical Perspective and Recent Research" Proceedings of the IEEE LXIV March 1976 Number 3, 329-354. [19]

Monday, 16 August 1976

Scientology's Guardian Office has authority over all Scientology legal actions, and is directing the FOIA cases against NSA, CIA, and other U.S. agencies and departments

Scientology's FCDC files suit in District Court to compel NSA to conduct a renewed search of

its files, and to enjoin NSA from any withholding of the materials. FCDC serves numerous interrogatories on NSA inquiring into its efforts to locate the records, its classification of documents, and its correspondence with CIA with respect to the NSA items that had been uncovered in FOIA actions against CIA. NSA declines to supply more than minimal information in answer to the interrogatories.

[NOTE: All Scientology FOIA actions are being handled by Scientology's Guardian Office, regardless of the specific Scientology organization filing the requests or suits.]

Tuesday, 24 August 1976

Opening Day of Scientology's "First International Conference for World Peace and Social Reform and Human Rights Prayer Day" at the Anaheim Convention Center in Anaheim, California. Ingo Swann has been promoted to be one of the many speakers at the event. Another one of the listed speakers is an unnamed "former Executive Assistant to the Deputy Director of the CIA." Also featured at the event are the Hubbard children—Diana, Suzette, Quentin, and Arthur—as well as Celebrity Centre director Yvonne (Gillham) Jentzsch. [3] [105]

Thursday, 28 October 1976

L-to-R: Diana, Quentin, Suzette, and Arthur Hubbard at the Human Rights Prayer Day event in Los Angeles just a little over two months before Quentin is found in Las Vegas in a coma from which he never recovers

Quentin Hubbard, son of L. Ron Hubbard, is found in a coma in a car parked near McCarren airport in Las Vegas, Nevada without any identification on him. He never comes out of the coma and dies just over two weeks later, still unidentified. Clark County Medical Examiner Sheldon Green determines the cause of death to be carbon monoxide poisoning, but says the "mode and manner" of death are unknown. Although ultimately able to identify Quentin through automobile records, the effort isn't made until after he has died. Autopsy reveals evidence of staph, and an angiogram had revealed a "possible cerebral abscess."

[NOTE: A little over a year later, Yvonne (Gillham) Jentzsch will die mysteriously, first diagnosed as having a staph infection, but with cause of death later being attributed to "a brain tumor."]

November 1976

George H. W. Bush (Sr.) is Director of Central Intelligence. He learns that Soviet officials have been visiting and questioning Hal Puthoff and Russell Targ at SRI. Bush requests and receives a briefing on CIA's investigations into parapsychology. Before there is any official response from Bush, he leaves CIA. [19]

1977

Monday, 31 January 1977

Remote Brain Targeting

As Director of CIA, George H.W. Bush is over CIA's remote viewing program while the Guardian Office is suing CIA for FOIA documents

Scientology's FCDC files a suit against the Director of CIA and others, No. 77-0175. The suit alleges that Scientology has been the subject of a government-wide conspiracy to destroy a religion. It claims that the church's constitutional and statutory rights have been violated in that the government agencies have improperly maintained and disseminated information; harassed, observed, and infiltrated the organization; "blacklisted" members; and subjected the organizations to discriminatory tax audits. Defendants include the Director of the FBI, the Attorney General of the United States, the Director of the CIA, the Secretary of the Treasury, the Chief of the National Central Bureau of the International Criminal Police Organization (INTERPOL); the Director of NSA; the Secretary of the Army; and the Postmaster General of the Postal Service. The United States is also named as a defendant.

[NOTE: George H.W. Bush leaves as Director of CIA at almost the same time this case is filed. This suit, as all the other Scientology FOIA cases, is being handled by Scientology's Guardian Office, run by Mary Sue Hubbard. Just over six months after this suit is filed, the Guardian Office is raided by the FBI and all of its senior members are charged with stealing IRS and other government agency documents. They will be sentenced to jail terms. In the aftermath, IRS's Meade Emory will tear down the entire Scientology corporate structure and rebuild it, but Meade Emory's work will be in secret, and the restructuring will be publicly attributed to L. Ron Hubbard, whose whereabouts are unknown the entire time.]

April 1977

The CIA's Office of Scientific Investigation completes a study about Soviet military and KGB applied parapsychology: "T. Hamilton; LSD/OSI; "Soviet and East European Parapsychology Research," SI 77-10012, April 1977 (SECRET/NOFORN)."[19]

Thursday, 2 June 1977

The United States District Court for the Central District of California issues a Summary Judgment for the U.S. in the Scientology FOIA action No. CV-75-3056-F (CSC v. Army), granting the Department of the Army the right withhold documents and portions of documents pertaining to Scientology and its founder, L. Ron Hubbard. Grounds are "national security."

On the same day, the United States District Court for the Central District of California issues a Summary Judgment for the U.S. in the Scientology FOIA action No. CV-75-4072-F (CSC v. Defense), granting the Department of Defense the right withhold documents and portions of documents pertaining to Scientology and its founder, L. Ron Hubbard. Grounds are "national security."

[NOTE: The Guardian Office still has FOIA actions outstanding against NSA, CIA, et al. Just over a month after this ruling, though, the Guardian Office will be raided by the FBI.]

Friday, 8 July 1977

Remote Brain Targeting

The FBI launches simultaneous early-morning raids on three Guardian Office locations: two in Los Angeles, one in Washington, D.C.

the Cedars complex and Fifield Manor in California, and the Washington, D.C. At the time of these raids, the Guardian Office is managing Freedom of Information Act (FOIA) suits against the Directors of CIA and the FBI, plus the CIA itself, the National Security Agency (NSA), the Department of Defense, Army Intelligence, Naval Intelligence, the Treasury Department (including IRS), INTERPOL, and the Attorney General of the United States.

[NOTE: The senior Guardian Office (GO) personnel will be sent to jail as a result of the raid, with the GO soon being disestablished altogether. Control over the other Scientology FOIA actions is severely compromised, and all are ultimately lost on grounds of "national security." Soon after, Meade Emory begins restructuring Scientology to have the ownership and control of the materials put under three non-Scientologist tax and probate attorneys. See 28 May 1982.]

The CIA has appropriated Scientology Advanced Technology via the use of Scientology OTs who have developed "remote-viewing" techniques, and who have trained CIA personnel. The CIA has a super-secret remote-viewing installation now set up, which has been joined to the U.S. Army Intelligence Agency's merger with the Army Security Agency to form the all-in-one "Intelligence and Security Command" (INSCOM). Under INSCOM, a major and super-secret remote-viewing program is being established at Fort Meade. It has been described thus: "The...researchers, in rivalry with their Soviet counterparts, were attempting nothing less than the development of the perfect spies, human beings who, undetectably and at almost zero cost, could spy upon the most remote, sensitive, and heavily guarded locations." The program has gone under the code name SCANATE (for "SCAN by coordiNATE"), but soon will become Project GRILLFLAME, and evolve into Project STAR GATE. The CIA, NSA, and the Defense Intelligence agencies are all fighting the Guardian Office FOIA actions, largely on the grounds of "national security" (although other justifications are thrown in for window dressing). Even Congress, other than the oversight committee, does not know about these secret intelligence projects utilizing Scientology and Scientologists.

Tuesday, 9 August 1977

CIA Director Stansfield Turner reveals publicly, but obliquely, that CIA has had "operational interest in parapsychology." [106] [19]

Friday, 4 November 1977

Former CIA Director Richard Helms appears in federal court in Washington, D.C. for sentencing on perjury before a Congressional Committee. Judge Barrington D. Parker reads Helm a stern lecture and announces the sentence: a $2000 fine and two years in jail—then suspends the sentence. [98]

1978

Tuesday, 17 January 1978

Founder of Celebrity Centre Yvonne (Gillham) Jentzsch dies under mysterious circumstances with similarities to medical findings in Quentin Hubbard's untimely death (Portrait by William Shirley)

Yvonne (Gillham) Jentzsch, who founded Celebrity Centre and has been closely connected to Hal Puthoff and Ingo Swann, dies at the Scientology Flag Land Base in Florida. There is a good deal of mystery surrounding her death. She had gone to Flag to be handled for a staph infection that had started in a leg, but then her death is reported as having been from a brain tumor.

May 1978

The SRI remote-viewing team is called upon to rapidly try and locate, with remote viewing, a downed Soviet Tupolev-22 bomber that had been configured for gathering electronic and photographic intelligence, and had gone down in the jungles the day before somewhere in Zaire. The task is given to two remote-viewers: Gary Langford at SRI (under Scientology OT VII Hal Puthoff), and a woman named Frances Bryan at Wright-Patterson Air Force Base. Both produce sketches of a river, which get matched to maps of the general area where the plane is thought to have been. A cabled summary of their results goes to the CIA station chief in Kinshasa, but the co-ordinates are over 70 miles from where the local CIA team believe the plane has gone down. The wreckage of the plane is soon found less than three miles from where the remote-viewers had pin-pointed it. CIA Director Stansfield Turner briefs President Jimmy Carter on the successful operation and recovery.

[NOTE: Seventeen years later, Carter briefly describes this incident during a speech at a college, even though the entire program is still top secret at the time.] [94]

Tuesday, 31 October 1978

Two senior scientists from the Soviet Ministry of Defense—Jan I. Koltunov and Nikolai A. Nosov—become members of the Moscow Bio-Electronics Laboratory, which is doing research in telepathy. [17]

Thursday, 7 December 1978

The KGB restructures the Moscow Laboratory for Bio-Electronics' "Rules for Admittance to Membership in the Central Public Laboratory for Bio-Electronics," creating stringent security requirements. [17]

1979

January 1979

Funding and tasking of remote viewing are being coordinated by the DIA, and the separate elements of the project are going by the collective code name GRILL FLAME. Integration of the project provides political cover for other agencies that might have been embarrassed to fund psychic spying directly. Atop this cover is Jack Vorona, who heads the DIA's Scientific and Technical Intelligence Directorate (known as "DT") as one of the Pentagon's top scientists. Funding for the SRI branch of the remote viewing operation alone, where Scientology OTs Hal Puthoff and Ingo Swann are operating, is estimated to run close to $1 million annually. [94]

Monday, 5 March 1979

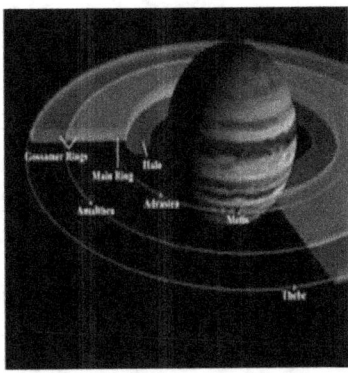

Jupiter is discovered to have rings

OT VII Hal Puthoff receives a call from the Jet Propulsion Laboratory (JPL) at La Canada, California. Raw data is coming in from the space probe Voyager 1, which is approaching Jupiter. JPL scientists have been completely surprised to discover that there is a ring around Jupiter. Puthoff's remote-viewing associate at SRI, OT VII Ingo Swann, had, on 27 April 1973 [see]–nearly six years earlier–remote-viewed Jupiter and had described and sketched just such a ring around the planet. Swann's results regarding Jupiter had been laughed off at the time. [107]

April 1979

OT VII Ingo Swann is listed in Scientology's Source magazine issue 20 as having completed New Era Dianetics for OTs.

Despite strict Scientology policies against it, Ingo Swann, while working for the CIA, is still doing upper-level Scientology services—even though Swann later claims he left Scientology "in 1975."

Sunday, 15 April 1979

OT VII Hal Puthoff issues an SRI Internal Report, "Feasibility Study on the Vulnerability of the MPS System to RV [Remote Viewing] Detection Techniques." [NOTE: MPS = Mapping, Chart, and Graphics Production System of the National Imagery and Mapping Agency (NIMA)] [108]

July 1979

Hella Hammid, a remote viewer working in the CIA-initiated program at SRI under OT VIIs Hal Puthoff and Ingo Swann, successfully describes microscopic picture targets as small as one millimeter square in an experimental series, and also correctly identifies a silver pin and a spool of thread inside an aluminum film can. [109] [110]

September 1979

The GRILL FLAME remote viewing headquarters at Fort Meade, Maryland is an outgrowth of the Scientology-based CIA-initiated remote viewing studies conducted at SRI by Scientology OT VIIs Hal Puthoff—who is Director of the SRI facility—and Ingo Swann. The Fort Meade unit is housed in two single-story wooden structures numbered 2560 and 2561. Fort Meade is a base for the National Security Agency (NSA) and part of the Army's Intelligence and Security Command (INSCOM), under which GRILL FLAME is officially established. GRILL FLAME takes its orders from the Pentagon's Office of the Army's Assistant Chief of Staff for Intelligence, and its tasking originates from CIA, DIA, and the President's National Security Council (NSC). Only a few dozen officials in the intelligence community have been briefed on the existence of GRILL FLAME. "Access is limited," an Army memorandum of the time notes, "to those personnel approved on a 'by name' basis."

Joseph McMoneagle is a consultant for the SRI remote viewing labs OT VII Hal Puthoff is Director, and where OT VII Ingo Swann trains government remote viewers. McMoneagle is being assigned numerous remote viewing tasks, for which he will later be granted a Legion of Merit award for excellence in intelligence service. *U.S. Intelligence agencies have become aware that the Russians have built the largest building under a single roof in the world. No one in the agencies, however, knows what is going on inside. The President's National Security Council staff orders INSCOM to have remote viewers see what they can determine about it. One of INSCOM's better remote-viewers, Joseph McMoneagle (a consultant with OT VII Hal Puthoff) reports, after his remote viewing of the facility, that a very large, new submarine with 18-20 missile launch tubes and a "large flat area" at the aft end will be launched in 100 days. Two Soviet subs, one with 24 launch tubes, and the other with 20 launch tubes and a large flat aft deck, are sighted 120 days later. These are new Soviet "Typhoon"-class submarines—the largest in the world. [111] [112] [113] [94]

Friday, 23 November 1979

The Joint Chiefs of Staff issue orders for Scientology-trained government remote-viewing personnel to begin providing information on the location and condition of the Iranian hostages.

[NOTE: A total of 206 remote-viewing sessions are ultimately devoted to the Iranian hostage crisis.]

1980s to Present

Friday, 28 May 1982

On the tenth anniversary of the purported Watergate first break in, a corporation called Church of Spiritual Technology (CST), doing business as the "L. Ron Hubbard Library," is created that controls all of L. Ron Hubbard's intellectual property, including all research and materials of Scientology, including the OT Levels. Three non-Scientology lawyers create the corporation and appoint themselves for life as its "Special Directors," vesting in themselves ultimate control over the corporation and all of the intellectual properties. The corporation has been created as part of a Scientology "restructuring" engineered by a former Assistant to Commissioner of IRS, Meade Emory.

Circa July 1982

OT VII Ingo Swann, under the direction of OT VII Harold Puthoff, head of the Remote Viewing Laboratory at SRI, is training remote viewers for the the Army. According to Major Ed Dames, he and five others are sent to be trained by Swann, purportedly in a "new model" of remote viewing.

[NOTE: Ed Dames has been documented as lying publicly about the involvement of Scientology and of OT VIIs Hal Puthoff and Ingo Swann in the genesis of remote viewing, and reasonably is viewed as a primary source of CIA disinformation and phony "technology" related to the subject. See CST and the CIA.]

Tuesday, 16 July 1984

A press release promises construction in Los Angeles of an $8 million "L. Ron Hubbard Library" where the original works of Hubbard "will be made readily accessible" to all. The release goes on to say that "until construction of the library," all the original manuscripts and tapes have been buried "in a series of underground vaults in half a dozen separate, but undisclosed locations."

[NOTE: No such library ever was built. The original works still remain buried in one or more underground vaults, of which only three have ever been identified. Also, the release omits any mention of the Meade Emory-created CST, which controls all the works, or of the fact that it is doing business at the time of the release under the exact name as the promised library (see 28 May 1982).] [114] [115]

Monday, 24 August 1992

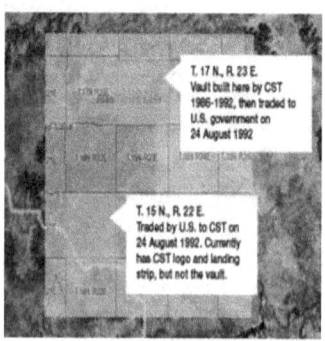

CST trades land with an underground vault to U.S. government at a major loss, including millions of dollars invested in building the vault. The land CST takes in trade is valued at only a little over $28,000.

The Church of Spiritual Technology (CST), which had been set up by a former Assistant to the Commissioner if IRS, Meade Emory, to control all of L. Ron Hubbard's works, makes a land swap with the U.S. government, giving the federal government one of the vaults it has constructed—the Trementina Base in New Mexico, and all the developments on it—in exchange for a like-sized but much cheaper piece of public and undeveloped land nearby. There is no accounting for the contents of the vault traded to the U.S. government. [NOTE: See Trementina Base for full coverage.]

Friday, 1 October 1993

Just over a year after the land swap (see 24 August 1992), the U.S. government restores tax exemption to Scientology.

[NOTE: See 1967 for date of revocation.]

The "Church of Spiritual Technology" owns all of Hubbard's works, and has buried the originals in one or more underground vaults—then traded one of the vaults to the U.S. government, contents unknown.

Wednesday, 6 September 1995

Ordered to declassify certain information about remote viewing, the CIA has its public relations office issue the following: "As mandated by Congress, CIA is reviewing available information and past research programs concerning parapsychological phenomena, mainly 'Remote Viewing' to determine whether they might have any utility for intelligence collection. CIA sponsored research on this subject in the 1970s. At that time, the program, always considered speculative and controversial, was determined to be unpromising." [116]

June 1998

Remote viewer Joe McMoneagle says in an interview about CIA's remote viewing programs: "Probably less than two percent of the information pertinent to the program has been released; certainly almost none of the operational data. A great deal of the research data is still classified as well." [117]

Afterword

As is well and thoroughly covered elsewhere, the "Church of Spiritual Technology" (CST) created altered versions of all of Hubbard's books and materials and began systematically replacing the originals with the altered versions. Even Hubbard taped lectures were edited, sometimes with entire sections removed. Earlier versions of the Hubbard works were collected up and destroyed. The claim was that the new versions were "correcting" the earlier versions.

Around the same time that the Guardian Office was destroyed, copies of what were purported to be the confidential upper materials began to be published in several media sources, first in a small Las Vegas rag called the Las Vegas Review-Journal, later in some broader publication magazines, and even excerpts in the Los Angeles Times and the Washington Post–whose Office of Naval Intelligence officer/reporter Bob Woodward told the world what it should be allowed to know about Watergate. No sources have ever been revealed for these purported confidential Scientology materials. Given the proven track of CST altering the works, and the fact that federal agencies had confidential upper-level Scientology materials in their files, there is sound foundation for the belief that the "OT Levels" in circulation are altered forgeries.

CST has made certain that it can never been proven one way or another by burying the original works in underground vaults, at least one of which they traded into the possession and control of the U.S. federal government on 24 August 1992.

In a nutshell, this research suggests that U.S. intelligence agencies recognized the power and potential of L. Ron Hubbard's work early on. They initially tried to work with Hubbard and Scientology to forward their secret agendas. When it eventually became clear that Hubbard would not cooperate, and in fact did his best to thwart these efforts, an intensive infiltration of Scientology was begun that gradually, yet dramatically changed the course of Scientology, such that it eventually diverted far from its original goals. For an intriguing documentary on remote viewing,

REFERENCES

1. Hunt, E. Howard Undercover, Memoirs of an American Secret Agent Berkely

2. Wells, Tom Wild Man; The Life and Times of Daniel Ellsberg

3. What is Scientology? Bridge Publications Los Angeles

4. Hubbard, L. Ron "The Story of Dianetics and Scientology" lecture of 18 October 1958

5. "Key Events in CIA's History," CIA Factbook on Intelligence 2002

6. Weiner, Tim "Robert Komer, 78, Figure in Vietnam, Dies" The New York Times 12 April 2000

7. Scan of Letter

8. Chase, Alston Harvard and the Unabomber: The Education of an American Terrorist W.W. Norton & Company 2003

9. Ross, Colin Bluebird: Deliberate Creation of Multiple Personality by Psychiatrists, Manitou Communications, 2000

10. Hubbard, L. Ron Dianetics, the Modern Science of Mental Health 1950

11. What is Scientology

12. Scan of Hubbard letter of resignation, 27 May 1950

13. Hubbard, L. Ron "How We Have Addressed the Problem of the Mind" taped lecture 4 July 1957

14. Hubbard, L. Ron "Group Dianetics" Dianetic Auditor's Bulletin Vol. 1 No. 7, January 1951

15. Hubbard, L. Ron Science of Survival limited manuscript edition Wichita, Kansas 25 June 1951

16. Project ARTICHOKE

17. Martin "Amplified Mind Power Research in The Former Socient Union" Retrieved

18. April 29, 2006

19. Hubbard, L. Ron What to Audit Scientific Press, Phoenix, Arizona July 1952 and "History of Man" Hubbard Association of Scientologists, London, July 1952

20. Kress, Dr. Kenneth A. "Parapsychology in Intelligence: A Personal Review and Conclusions" Studies in Intelligence (CIA publication) Winter 1977

21. Schwalbe, David "LSD and the CIA, Part 2" Dateline 14 March 1999

22. Hubbard, L. Ron Philadelphia Doctorate Course lecture series

23. Kutler, Stanley I. Abuse of Power: the New Nixon Tapes

24. Lee, Martin A. and Shlain, Bruce Acid Dreams; The CIA, LSD and the Sixties Rebellion Grove Press, New York: 1985

25. MK-ULTRA

26. Martin, Harry V. and Caul, David "Mind Control" Napa Sentinel August-November 1991 1

27. Hubbard, L. Ron "Politics, Freedom From" LRH Secretarial Executive Directive 56 Int 14 June 1965 reissued as Hubbard Communication Office Policy Letter 10 January 1968

28. Declassified Documents—Microfilms Under MKULTRA" Research Publications Woodbridge, CT 1984 002258

29. Church of Scientology vs. Commissioner of Internal Revenue Docket No. 3352-78 United States Tax Court filed 24 September 1984

30. Miller, Russell Bare Faced Messiah

31. Ellsberg, Daniel Secrets: A Memoir of Vietnam and the Pentagon Papers Penguin 2003

32. "Plants Do Worry and Feel Pain," Garden News 18 December 1959

33. Hubbard, L. Ron "Creation and Goals" a recorded lecture of 3 August 1961

34. Sea Org Orders of the Day (OODs) 28 February 1969

35. Judiciary Committee Impeachment hearings, Testimony of Witnesses, Book III: Responses by CIA to questions submitted by the Committee

36. Cooper, Paulette The Scandal of Scientology

37. Hubbard, L. Ron "Level VII" a taped lecture of 23 February 1965

38. Hubbard, L. Ron "The Well Rounded Auditor" a taped lecture of 29 June 1965

39. A.J. Weberman

40. Hubbard, L. Ron, Scientology Policy of 28 December 1965, revised 1968, "Enrollment in Suppressive Groups"

41. Hubbard, L. Ron, Scientology Policy of 6 December 1976, revised 8 April 1988, "Illegal PCs, Acceptance Of"

42. Burton, Christine "Green Music" article

43. Hubbard, L. Ron "Clearing Course Security" Scientology policy letter of 16 August 1966

44. Hubbard, L. Ron "OT Personnel" Scientology policy letter 10 November 1966 Issue II

45. Swann, Ingo Remote Viewing – The Real Story

46. Miller, Russel Interview with David Mayo

47. Miller, Russell Bare Faced Messiah

48. Meade Emory profile

49. Tanner, Jerald "Mormon Spies, Hughes and the CIA" citing testimony before Judiciary Committee Impeachment hearings, Book III

50. CIA memo #104-10119-10323 from CIA Chief Central Cover Staff Corporate Cover Branch

51. James McCord biography

52. Liddy, G. Gordon Will, the Autobiography of G. Gordon Liddy St. Martin's

53. Hal, Puthoff "Success Story" Scientology Advanced Org Los Angeles (AOLA) special publication, blue painting cover, printed in 1971

54. Liddy, G. Gordon Deposition in Dean v. Liddy et al., U.S. District Court D.C. 92-1807

55. Smith, J. "List of CIA Agents" Intelligence/Parapolitics magazine Brussels November 1985

56. Miller, Russell Interview with Kima Douglas

57. "Data Concerning the Death of Scientology Parishioner Susan Meister" Company Memorandum TSMY Apollo

58. White House Plumbers

59. Campaign Contributions Task Force #804 – Hughes / Rebozo Investigations, Box 86, Caulfield, John: "7/71 Sandwedge proposal"

60. Caulfied, John J. (Jack Caulfield) "In Their Own Words"

61. Memorandum for the Record: "Summary of Mr. Karl Wagner's Knowledge of CIA Assistance to Mr. E. Howard Hunt" Judiciary Committee Impeachment hearings

62. Memorandum for the Record: "Summary of Contacts by Mr. Stephen Carter Greenwood with Mr. E. Howard Hunt" Judiciary Committee Impeachment hearings Book III

63. "Orders of the Day" (OODs) of the Scientology Flagship Apollo 1971-1972

64. Hubbard, L. Ron "Advanced Courses" Scientology policy letter of 12 August 1971

65. Transcript of recording of a meeting among the President, John Dean, and H.R. Haldeman in the Oval Office on March 17, 1973 from 1:25 pm to 2:10 pm

66. Bernard Barker testimony, May 11 and May 24, 1973 Judiciary Committee Impeachment Hearings Book I, Events Prior to the Watergate Break-in

67. Victorian, Armen, quoting Ingo Swann "Remote Viewing and the U.S. Intelligence Community" Lobster Issue 31: June 1996

68. Brussell, Mae: "Why Was Martha Mitchell Kidnapped" The Realist August 1972

69. House of Representatives Judiciary Committee Impeachment Hearings, Book I

70. "Chasing George W. Bush and the F-102"

71. Citrine, Charlie Watergate Timeline

72. LaMother, Captain John D. Defense Intelligence Agency (DIA) Report: "Controlled Offensive Behavior—USSR" DIA Task Number T72-01-14 Controlled Offensive Behavior – USSR large PDF file

73. Caddy, Douglas "Gay Bashing and Watergate" Advocate.com 1 August 2005

74. FBI files on L. Ron Hubbard

75. United States Court of Appeals, Eighth Circuit ruling April 10, 1972 Church of Scientology of Minnesota et al. v. Department of Health, Education & Welfare, etc., et al. No. 71-1507

76. "Bug Suspects Got Campaign Funds" Washington Post

77. FBI report dated 22 June 1972, "Memorandum to Mr. Bolz"

78. Congressional testimony of Alfred Baldwin, 24 May 1973

79. Document in PDF format "BushGuardmay4.pdf" released by CBS news in September 2004

80. U.S. v. George Gordon Liddy et al. Grand Jury Indictment; Grand Jury sworn in on

81. **June 5, 1972**

82. James McCord testimony, May 11 and May 24, 1973 Judiciary Committee Impeachment Hearings Book I, Events Prior to the Watergate Break-in

83. Document in PDF format "BushGuardmay19.pdf" released by CBS news in September 2004

84. Excerpt of letter from Nixon to Kissinger and Haig

85. The Public Papers of President Richard Nixon; 1972

86. U.S. vs. G. Gordon Liddy, appellant No. 73-1565 United States Court of Appeals District of Columbia, decided 8 November 1974

87. Watergate first break in

88. Arlington National Cemetry web page on John Paul Vann

89. Helms, Richard and Hood, William A Look Over My Shoulder Random House, 2003

90. Testimony of L. Patrick Gray, former Acting Director of the Federal Bureau of Investigation, in Congressional hearings, 3 and 6 August 1973

91. Transcript from web site "History and Politics Out Loud"

92. Church of Scientology v. IRS, No. 3352-78, United States Tax Court, filed 24 September 1978

93. Watergate Chronology

94. Watergate burglars indicted" NBC News abstract

95. House of Representatives Judiciary Committee Impeachment Hearings, Book III

96. Schnabel, Jim Remote Viewers: The Secret History of America's Psychic Spies Dell (1997)

97. Bio of Jack Caulfield

98. "Project MKULTRA, the CIA's Program of Research in Behavioral Modification" Report by U.S. Senate Select Committee on Intelligence

99. The New York Times Company Timeline: NY Times Timeline 1971-2000

100. Powers, Thomas "Inside the Department of Dirty Tricks" Atlantic Monthly August 1979 Volume 244 No. 2 pages 33-64

101. "Pentagon Papers: Case Dismissed" Time magazine 21 May 1973

102. "Break-In Memo Sent to Ehrlichman" Washington Post Staff Writers Wednesday, June 13, 1973

103. Jensen, Derrick "The Plants Respond: An Interview with Cleve Backster" The Sun July 1997

104. Celebrity magazine Minor Issue 8 November 1973

105. RTC v. FACTnet, Inc. US District Court Colorado No. 95B2143 testimony of Robert Vaughn Young 21 September 1995

106. Celebrity magazine Minor Issue 11 September 1974

107. Celebrity magazine Major Issue 21

108. O'Leary, J. "Turner Denies CIA Bugging of South Korea's Park" The Washington Star 9 August 1977.

109. Swann, Ingo the 1973 Remote Viewing Probe of the Planet Jupiter

110. Puthoff, Hal "CIA-Initiated RV Program at SRI" article

111. Puthoff, Harold and Targ, Russell, "Direct Perception of Remote Geographical Locations", SRI Menlo Park, 1979

112. Targ, Russell Miracles of Mind; Remote Viewing

113. May, Dr. Edwin C. "Response to the CIA/AIR Report on Remote Viewing"

114. "Interview with Joseph McMoneagle" Psychic World Summer issue 1998

115. STAR GATE (Controlled Remote Viewing)

116. Untitled PR Newswire press release dated July 17 1984 but with July 16 dateline, begins "Construction of an $8 million library..."

117. Untitled PR Newswire press release dated July 17 1984 with July 17 dateline, begins "Construction of an $8 million library..." (text different from similar release with 16 July 1984 dateline)

118. CIA Public Affairs Office "CIA Statement on 'Remote Viewing'", 6 September 1995

119. Csere, Tom "Interview with Joe McMoneagle, World-Class Remote Viewer" Psychic World Summer 1998

120. Colodny, Len and Gettlin, Robert Silent Coup: The Removal of a President

CHAPTER TWO

"Power is in tearing human minds to pieces and putting them together again in new shapes of your own choosing."

- George Orwell (1984)

Herbert Spencer first used the phrase – after reading Charles Darwin's On the Origin of Species – in his Principles of Biology (1864), in which he drew parallels between his own economic theories and Darwin's biological ones, writing, "This survival of the fittest, which I have here sought to express in mechanical terms, is that which Mr. Darwin has called 'natural selection', or the preservation of favored races in the struggle for life."

Dick Sutphen is a Spiritual Healing, Hypnotherapy and Human Potential. He is President of the American Board of Hypnotherapy, and the author of the million-copy reincarnation bestseller, You Were Born Again to Be Together (Simon and Schuster Pocket Books). His latest book is Soul Agreements (Hampton Roads Publishing). As a specialist in brain/mind technology, Dick instructs medical professionals in how to use his life-changing techniques. He has created over 300 mind-programming CDs. Over 180,000 people have attended a Sutphen Seminar, workshop, or retreat. Dick is also a featured speaker and teacher at all the major Professional Hypnosis conferences. He has appeared on over 400 radio and television shows.

The Battle for Your Mind: Brainwashing Techniques Being used On the Public by Dick Sutphen

SUMMARY OF CONTENTS

1. The Birth of Conversion

2. Brainwashing in Christian Revivalism in 1735

3. The Pavlovian explanation of the three brain phases

4. Born-again preachers: Step-by-Step, how they conduct a revival and the expected physiological results

5. The "voice roll" technique used by preachers, lawyers and hypnotists

6. New Trance-Inducing Churches - The 6 steps to conversion.

7. The Decognition Process

8. Thought-Stopping Techniques

9. The "Sell it by Zealot" Technique

10. True Believers and Mass Movements

11. Persuasion Techniques: "Yes set," "Imbedded Commands," "Shock and Confusion, "and the "Interspersal Technique"

12. Subliminals

13. Vibrato and ELF waves

14. Inducing Trance with Vibrational Sound - Even professional observers will be "possessed" at charismatic gatherings

15. The "Only Hope" Technique to attend and not be converted

16. Non-Detectable Neurophone programming through the skin

17. The Medium for Mass Take-over

Dick Sutphen's tape is a studio-recorded, expanded version of a talk he delivered at the World Congress of Professional Hypnotists Convention in Las Vegas, Nevada. Although the tape carries a copyright to protect it from unlawful duplication for sale by other companies, in this case Mr. Sutphen has invited individuals to make copies and give them to friends or anyone in a position to communicate this information.

He further states, "Although I've been interviewed about the subject on many local and regional radio and TV talk shows, large-scale mass communication appears to be blocked, since it could result in suspicion or investigation of the very media presenting it or the sponsors that support the media. Some government agencies do not want this information generally known. Nor do the Born-Again Christian movement, cults, and many human-potential trainings."

"Everything I will relate only exposes the surface of the problem. I don't know how the misuse of these techniques can be stopped. I don't think it is possible to legislate against that which often cannot be detected; and if those who legislate are using these techniques, there is little hope of affecting laws to govern usage. I do know that the first step to initiate change is to generate interest. In this case, that will probably only result from an underground effort."

In talking about this subject, I am talking about my own business. I know it, and I know how effective it can be. I produce hypnosis and subliminal tapes and, in some of my seminars, I use conversion tactics to assist participants to become independent and self-sufficient. But, anytime I use these techniques, I point out that I am using them, and those attending have a choice to participate or not. They also know what the desired result of participation will be.

So, to begin, I want to state the most basic of all facts about brainwashing: In the entire history of man, no one has ever been brainwashed and realized, or believed, that he had been brainwashed. Those who have been brainwashed will usually passionately defend their manipulators, claiming they have simply been "shown the light" ... or have been transformed in miraculous ways.

The Birth of Conversion

CONVERSION is a "nice" word for BRAINWASHING and any study of brainwashing has to begin with a study of Christian revivalism in eighteenth century America. Apparently, Jonathan Edwards accidentally discovered the techniques during a religious crusade in 1735 in Northampton, Massachusetts. By inducing guilt and acute apprehension and by increasing the tension, the "sinners" attending his revival meetings would break down and completely submit. Technically, what Edwards was doing was creating conditions that wipe the brain slate clean so that the mind accepts new programming. The problem was that the new input was negative. He would tell them, "You're a sinner! You're destined for hell!"

As a result, one person committed suicide and another attempted suicide. And the neighbors of the suicidal converts related that they, too, were affected so deeply that, although they had found "eternal salvation," they were obsessed with a diabolical temptation to end their own lives.

Once a preacher, cult leader, manipulator or authority figure creates the brain phase to wipe the brain-slate clean, his subjects are wide open. New input, in the form of suggestion, can be substituted for their previous ideas. Because Edwards didn't turn his message positive until the end of the revival, many accepted the negative suggestions and acted, or desired to act, upon them.

Charles J. Finney was another Christian revivalist who used the same techniques four years later in mass religious conversions in New York. The techniques are still being used today by Christian revivalists, cults, human-potential trainings, some business rallies, and the United States Armed Services. This is to name just a few.

Let me point out here that I don't think most revivalist preachers realize or know they are using brainwashing techniques. Edwards simply stumbled upon a technique that really worked, and others copied it and have continued to copy it for over two hundred years.

And the more sophisticated our knowledge and technology become, the more effective the conversion. I feel strongly that this is one of the major reasons for the increasing rise in Christian fundamentalism, especially the televised variety, while most of the orthodox religions are declining.

The Three Brain Phases

The Christians may have been the first to successfully formulate brainwashing, but we have to look to Pavlov, the Russian scientist, for a technical explanation. In the early 1900s, his work with animals opened the door to further investigations with humans. After the revolution in Russia, Lenin was quick to see the potential of applying Pavlov's research to his own ends.

Three distinct and progressive states of trans-marginal inhibition were identified by Pavlov. The first is the EQUIVALENT phase, in which the brain gives the same response to both strong and weak stimuli. The second is the PARADOXICAL phase, in which the brain responds more actively to weak stimuli than to strong. And the third is the ULTRA- PARADOXICAL phase, in which conditioned responses and behavior patterns turn from positive to negative or from negative to positive.

With the progression through each phase, the degree of conversion becomes more effective and complete. The way to achieve conversion are many and varied, but the usual first step in religious or political brainwashing is to work on the emotions of an individual or group until they reach an abnormal level of anger, fear, excitement, or nervous tension.

The progressive result of this mental condition is to impair judgment and increase suggestibility. The more this condition can be maintained or intensified, the more it compounds. Once catharsis, or the first brain phase, is reached, the complete mental takeover becomes easier. Existing mental programming can be replaced with new patterns of thinking and behavior.

Other often-used physiological weapons to modify normal brain functions are fasting, radical or high sugar diets, physical discomforts, regulation of breathing, mantra chanting in meditation, the disclosure of awesome mysteries, special lighting and sound effects, programed response to incense, or intoxicating drugs.

The same results can be obtained in contemporary psychiatric treatment by electric shock treatments and even by purposely lowering a person's blood sugar level with insulin injections.

Before I talk about exactly how some of the techniques are applied, I want to point out that hypnosis and conversion tactics are two distinctly different things, and that conversion techniques are far more powerful. However, the two are often mixed with powerful results.

How Revivalist Preachers Work

If you'd like to see a revivalist preacher at work, there are probably several in your city. Go to the church or tent early and sit in the rear, about three-quarters of the way back. Most likely repetitive music will be played while the people come in for the service. A repetitive beat, ideally ranging from 45 to 72 beats per minute (a rhythm close to the beat of the human heart), is very hypnotic and can generate an eyes-open altered state of consciousness in a very high percentage of people. And, once you are in an alpha state, you are at least 25 times as suggestible as you would be in full beta consciousness. The music is probably the same for every service, or incorporates the same beat, and many of the people will go into an altered state almost immediately upon entering the sanctuary. Subconsciously, they recall their state of mind from previous services and respond according to the post-hypnotic programming.

Watch the people waiting for the service to begin. Many will exhibit external signs of trance body relaxation and slightly dilated eyes. Often, they begin swaying back and forth with their hands in the air while sitting in their chairs. Next, the assistant pastor will probably come out. He usually speaks with a pretty good "voice roll."

Voice Roll Technique

A "voice roll" is a patterned, paced style used by hypnotists when inducing a trance. It is also used by many lawyers, several of whom are highly trained hypnotists, when they desire to entrench a point firmly in the minds of the jurors. A voice roll can sound as if the speaker was talking to the beat of a metronome or it may sound as though he were emphasizing every word in a monotonous, patterned style. The words will usually be delivered at the rate of 45 to 60 beats per minute, maximizing the hypnotic effect.

Now the assistant pastor begins the "build-up" process. He induces an altered state of consciousness and/or begins to generate the excitement and the expectations of the audience. Next, a group of young women in "sweet and pure" chiffon dresses might come out to sing a song. Gospel songs are great for building excitement and INVOLVEMENT. In the middle of the song, one of the girls might be "smitten by the spirit" and fall down or react as if possessed by the Holy Spirit. This very effectively increases the intensity in the room. At this point, hypnosis and conversion tactics are being mixed. And the result is the audience's attention span is now totally focused upon the communication while the environment becomes more exciting or tense.

Right about this time, when an eyes-open mass-induced alpha mental state has been achieved, they will usually pass the collection plate or basket. In the background, a 45-beat-per-minute voice roll from the assistant preacher might exhort, "Give to God . . . Give to God . . . Give to God . . ." And the audience does give. God may not get the money, but his already wealthy representative will.

Next, the fire-and-brimstone preacher will come out. He induces fear and increases the tension by talking about "the devil," "going to hell," or the forthcoming Armageddon.

In the last such rally I attended, the preacher talked about the blood that would soon be running out of every faucet in the land. He was also obsessed with a "bloody axe of God," which everyone had seen hanging above the pulpit the previous week. I have no doubt that everyone saw it the power of suggestion given to hundreds of people in hypnosis assures that at least 10 to 25 percent would see whatever he suggested they see.

In most revivalist gatherings, "testifying" or "witnessing" usually follows the fear-based sermon. People from the audience come up on stage and relate their stories. "I was crippled and now I can walk!" "I had arthritis and now it's gone!" It is a psychological manipulation that works. After listening to numerous case histories of miraculous healings, the average guy in the audience with a minor problem is sure he can be healed. The room is charged with fear, guilt, intense excitement, and expectations.

Now those who want to be healed are frequently lined up around the edge of the room, or they are told to come down to the front. The preacher might touch them on the head firmly and scream, "Be healed!" This releases the psychic energy and, for many, catharsis results. Catharsis is a purging of repressed emotions. Individuals might cry, fall down or even go into spasms. And if catharsis is affected, they stand a chance of being healed. In catharsis (one of the three brain phases mentioned earlier), the brain-slate is temporarily wiped clean and the new suggestion is accepted.

For some, the healing may be permanent. For many, it will last four days to a week, which is, incidentally, how long a hypnotic suggestion given to a somnambulistic subject will usually last. Even if the healing doesn't last, if they come back every week, the power of suggestion may continually override the problem or sometimes, sadly, it can mask a physical problem which could prove to be very detrimental to the individual in the long run.

I'm not saying that legitimate healings do not take place. They do. Maybe the individual was ready to let go of the negativity that caused the problem in the first place; maybe it was the work of God. Yet I contend that it can be explained with existing knowledge of brain/mind function.

The techniques and staging will vary from church to church. Many use "speaking in tongues" to generate catharsis in some while the spectacle creates intense excitement in the observers.

The use of hypnotic techniques by religions is sophisticated, and professionals are assuring that they become even more effective. A man in Los Angeles is designing, building, and reworking a lot of churches around the country. He tells ministers what they need and how to use it. This man's track record indicates that the congregation and the monetary income will double if the minister follows his instructions. He admits that about 80 percent of his efforts are in the sound system and lighting.

Powerful sound and the proper use of lighting are of primary importance in inducing an altered state of consciousness. I've been using them for years in my own seminars. However, my participants are fully aware of the process and what they can expect as a result of their participation.

Six Conversion Techniques

Cults and human-potential organizations are always looking for new converts. To attain them, they must also create a brain-phase. And they often need to do it within a short space of time, a weekend, or maybe even a day. The following are the six primary techniques used to generate the conversion.

The meeting or training takes place in an area where participants are cut off from the outside world. This may be any place: a private home, a remote or rural setting, or even a hotel ballroom where the participants are allowed only limited bathroom usage.

In human-potential trainings, the controllers will give a lengthy talk about the importance of "keeping agreements" in life. The participants are told that if they don't keep agreements, their life will never work. It's a good idea to keep agreements, but the controllers are subverting a positive human value for selfish purposes. The participants vow to themselves and their trainer that they will keep their agreements. Anyone who does not will be intimidated into agreement or forced to leave. The next step is to agree to complete training, thus assuring a high percentage of conversions for the organizations. They will USUALLY have to agree not to take drugs, smoke, and sometimes not to eat or they are given such short meal breaks that it creates tension. The real reason for the agreements is to alter internal chemistry, which generates anxiety and hopefully causes at least a slight malfunction of the nervous system, which in turn increases the conversion potential.

Before the gathering is complete, the agreements will be used to ensure that the new converts go out and find new participants. They are intimidated into agreeing to do so before they leave. Since the importance of keeping agreements is so high on their priority list, the converts will twist the arms of everyone they know, attempting to talk them into attending a free introductory session offered at a future date by the organization. The new converts are zealots. In fact, the inside term for merchandising the largest and most successful human-potential training is, "sell it by zealot!"

At least a million people are graduates and a good percentage has been left with a mental activation button that assures their future loyalty and assistance if the guru figure or organization calls. Think about the potential political implications of hundreds of thousands of zealots programmed to campaign for their guru.

Be wary of an organization of this type that offers follow-up sessions after the seminar. Follow-up sessions might be weekly meetings or inexpensive seminars given on a regular basis which the organization will attempt to talk you into taking or any regularly scheduled event used to maintain control. As the early Christian revivalists found, long-term control is dependent upon a good follow-up system.

Alright, let's look at the second tip-off that indicates conversion tactics are being used. A schedule is maintained that causes physical and mental fatigue. This is primarily accomplished by long hours in which the participants are given no opportunity for relaxation or reflection.

The third tip-off: techniques used to increase the tension in the room or environment.

Number four: Uncertainty. I could spend hours relating various techniques to increase tension and generate uncertainty. Basically, the participants are concerned about being "put on the spot" or encountered by the trainers, guilt feelings are played upon, participants are tempted to verbally relate their innermost secrets to the other participants or forced to take part in activities that emphasize removing their masks. One of the most successful human-potential seminars forces the participants to stand on a stage in front of the entire audience while being verbally attacked by the trainers. A public opinion poll, conducted a few years ago, showed that the number one most-fearful situation an individual could encounter is to speak to an audience. It ranked above window washing outside the 85th floor of an office building. So, you can imagine the fear and tension this situation generates within the participants. Many faint, but most cope with the stress by mentally going away. They literally go into an alpha state, which automatically makes them many times as suggestible as they normally are. And another loop of the downward spiral into conversion is successfully affected.

The fifth clue that conversion tactics are being used is the introduction of jargon, new terms that have meaning only to the "insiders" who participate. Vicious language is also frequently used, purposely, to make participants uncomfortable.

The final tip-off is that there is no humor in the communications at least until the participants is converted. Then, merry-making and humor are highly desirable as symbols of the new joy the participants have supposedly "found."

I'm not saying that good does not result from participation in such gatherings. It can and does. But I contend it is important for people to know what has happened and to be aware that continual involvement may not be in their best interest.

Over the years, I've conducted professional seminars to teach people to be hypnotists, trainers, and counselors. I've had many of those who conduct trainings and rallies come to me and say, "I'm here because I know that what I'm doing works, but I don't know why." After showing them how and why, many have gotten out of the business or have decided to approach it differently or in a much more loving and supportive manner.

Many of these trainers have become personal friends, and it scares us all to have experienced the power of one person with a microphone and a room full of people. Add a little charisma and you can count on a high percentage of conversions. The sad truth is that a high percentage of people want to give away their power, they are true "believers"!

Cult gatherings or human-potential trainings are an ideal environment to observe first-hand what is technically called the "Stockholm Syndrome." This is a situation in which those who are intimidated, controlled, or made to suffer, begin to love, admire, and even sometimes sexually desire their controllers or captors.

But let me inject a word of warning here: If you think you can attend such gatherings and not be affected, you are probably wrong. A perfect example is the case of a woman who went to Haiti on a Guggenheim Fellowship to study Haitian Voodoo. In her report, she related how the music eventually induced uncontrollable bodily movement and an altered state of consciousness. Although she understood the process and thought herself above it, when she began to feel herself become vulnerable to the music, she attempted to fight it and turned away. Anger or resistance almost always assures conversion. A few moments later she was possessed by the music and began dancing in a trance around the Voodoo meeting house. A brain phase had been induced by the music and excitement, and she awoke feeling reborn. The only hope of attending such gatherings without being affected is to be a Buddha and allow no positive or negative emotions to surface. Few people are capable of such detachment.

Before I go on, let's go back to the six tip-offs to conversion. I want to mention the United States Government and military boot camp. The Marine Corps talks about breaking men down before "rebuilding" them as new men as marines! Well, that is exactly what they do, the same way a cult breaks its people down and rebuilds them as happy flower sellers on your local street corner. Every one of the six conversion techniques are used in boot camp. Considering the needs of the military, I'm not making a judgment as to whether that is good or bad. IT IS A FACT that the men are effectively brainwashed. Those who won't submit must be discharged or spend much of their time in the brig.

Decognition Process

Once the initial conversion is effected, cults, armed services, and similar groups cannot have cynicism among their members. Members must respond to commands and do as they are told; otherwise they are dangerous to the organizational control. This is normally accomplished in a three step Decognition Process.

Step one is ALERTNESS REDUCTION: The controllers cause the nervous system to malfunction, making it difficult to distinguish between fantasy and reality. This can be accomplished in several ways. POOR DIET is one; watch out for Brownies and Kool Aide. The sugar throws the nervous system off. More subtle is the "SPIRITUAL DIET" used by many cults. They eat only vegetables and fruits; without the grounding of grains, nuts, seeds, dairy products, fish or meat, an individual becomes mentally "spacey." INADEQUATE SLEEP is another primary way to reduce alertness, especially when combined with long hours of work or intense physical activity. Also, being bombarded with intense and unique experiences achieves the same result.

Step Two is PROGRAMED CONFUSION: You are mentally assaulted while your alertness is being reduced as in Step One. This is accomplished with a deluge of new information, lectures, discussion groups, encounters or one-to-one processing, which usually amounts to the controller bombarding the individual with questions. During this phase of decognition, reality and illusion often merge and perverted logic is likely to be accepted.

Step Three is THOUGHT STOPPING: Techniques are used to cause the mind to go "flat." These are altered-state-of-consciousness techniques that initially induce calmness by giving the mind something simple to deal with and focusing awareness. The continued use brings on a feeling of elation and eventually hallucination. The result is the reduction of thought and eventually, if used long enough, the cessation of all thought and withdrawal from everyone and everything except that which the controllers direct. The takeover is then complete. It is important to be aware that when members or participants are instructed to use "thought-stopping" techniques, they are told that they will benefit by so doing: they will become "better soldiers" or "find enlightenment."

There are three primary techniques used for thought stopping. The first is MARCHING: the thump, thump, thump beat literally generates self-hypnosis and thus great susceptibility to suggestion.

The second thought stopping technique is MEDITATION. If you spend an hour to an hour and a half a day in meditation, after a few weeks, there is a great probability that you will not return to full beta consciousness. You will remain in a fixed state of alpha for as long as you continue to meditate. I'm not saying this is bad - if you do it yourself. It may be very beneficial. But it is a fact that you are causing your mind to go flat. I've worked with meditators on an EEG machine and the results are conclusive: the more you meditate, the flatter your mind becomes until, eventually and especially if used to excess or in combination with decognition, all thought ceases. Some spiritual groups see this as nirvana - which is bullshit. It is simply a predictable physiological result. And if heaven on earth is non- thinking and non-involvement, I really question why we are here.

The third thought-stopping technique is CHANTING, and often chanting in meditation. "Speaking in tongues" could also be included in this category.

All three-stopping techniques produce an altered state of consciousness. This may be very good if YOU are controlling the process, for you also control the input. I personally use at least one self-hypnosis programming session every day and I know how beneficial it is for me. But you need to know if you use these techniques to the degree of remaining continually in alpha that, although you'll be very mellow, you'll also be more suggestible.

True Believers & Mass Movements

Before ending this section on conversion, I want to talk about the people who are most susceptible to it and about Mass Movements. I am convinced that at least a third of the population is what Eric Hoffer calls "true believers." They are joiners and followers . . . people who want to give away their power. They look for answers, meaning, and enlightenment outside themselves.

Hoffer, who wrote THE TRUE BELIEVER, a classic on mass movements, says, "true believers are not intent on bolstering and advancing a cherished self, but are those craving to be rid of unwanted self. They are followers, not because of a desire for self-advancement, but because it can satisfy their passion for self-renunciation!" Hoffer also says that true believers "are eternally incomplete and eternally insecure"!

I know this from my own experience. In my years of communicating concepts and conducting trainings, I have run into them again and again. All I can do is attempting to show them that the only thing to seek is the True Self within. Their personal answers are to be found there and there alone. I communicate that the basics of spirituality are self-responsibility and self-actualization. But most of the true believers just tell me that I'm not spiritual and go looking for someone who will give them the dogma and structure they desire. Never underestimate the potential danger of these people. They can easily be molded into fanatics who will gladly work and die for their holy cause. It is a substitute for their lost faith in them and offers them as a substitute for individual hope. The Moral Majority is made up of true believers. All cults are composed of true believers. You'll find them in politics, churches, businesses, and social cause groups. They are the fanatics in these organizations.

Mass Movements will usually have a charismatic leader. The followers want to convert others to their way of living or impose a new way of life if necessary, by legislating laws forcing others to their view, as evidenced by the activities of the Moral Majority. This means enforcement by guns or punishment, for that is the bottom line in law enforcement.

A common hatred, enemy, or devil is essential to the success of a mass movement. The Born-Again Christians have Satan himself, but that isn't enough they've added the occult, the New Age thinkers and, lately, all those who oppose their integration of church and politics, as evidenced in their political re-election campaigns against those who oppose their views. In revolutions, the devil is usually the ruling power or aristocracy. Some human-potential movements are far too clever to ask their graduates to join anything, thus labeling themselves as a cult but, if you look closely, you'll find that their devil is anyone and everyone who hasn't taken their training. There are mass movements without devils but they seldom attain major status. The True Believers are mentally unbalanced or insecure people, or those without hope or friends. People don't look for allies when they love, but they do when they hate or become obsessed with a cause. And those who desire a new life and a new order feel the old ways must be eliminated before the new order can be built.

Persuasion Techniques

Persuasion isn't technically brainwashing but it is the manipulation of the human mind by another individual, without the manipulated party being aware what caused his opinion shift. I only have time to very basically introduce you to a few of the thousands of techniques in use today, but the basis of persuasion is always to access your RIGHT BRAIN. The left half of your brain is analytical and rational. The right side is creative and imaginative. That is overly simplified but it makes my point. So, the idea is to distract the left brain and keep it busy. Ideally, the persuader generates an eyes-open altered state of consciousness, causing you to shift from beta awareness into alpha; this can be measured on an EEG machine.

First, let me give you an example of distracting the left brain. Politicians use these powerful techniques all the time; lawyers use many variations which, I've been told, they call "tightening the noose."

Assume for a moment that you are watching a politician give a speech. First, he might generate what is called a "YES SET." These are statements that will cause listeners to agree; they might even unknowingly nod their heads in agreement. Next comes the

TRUISMS. These are usually facts that could be debated but, once the politician has his audience agreeing, the odds are in the politician's favor that the audience won't stop to think for themselves, thus continuing to agree. Last comes the SUGGESTION. This is what the politician wants you to do and, since you have been agreeing all along, you could be persuaded to accept the suggestion. Now, if you'll listen closely to my political speech, you'll find that the first three are the "yes set," the next three are truisms and the last is the suggestion.

"Ladies and gentlemen: are you angry about high food prices? Are you tired of astronomical gas prices? Are you sick of out-of-control inflation? Well, you know the Other Party allowed 18 percent inflation last year; you know crime has increased 50 percent nationwide in the last 12 months, and you know your paycheck hardly covers your expenses any more. Well, the answer to resolving these problems is to elect me, John Jones, to the U.S. Senate."

And I think you've heard all that before. But you might also watch for what are called Imbedded Commands. As an example: On key words, the speaker would make a gesture with his left hand, which research has shown is more apt to access your right brain. Today's media-oriented politicians and spellbinders are often carefully trained by a whole new breed of specialist who are using every trick in the book, both old and new - to manipulate you into accepting their candidate.

The concepts and techniques of Neuro-Linguistics are so heavily protected that I found out the hard way that to even talk about them publicly or in print results in threatened legal action. Yet Neuro-Linguistic training is readily available to anyone willing to devote the time and pay the price. It is some of the most subtle and powerful manipulation I have yet been exposed to. A good friend who recently attended a two-week seminar on Neuro-Linguistics found that many of those she talked to during the breaks were government people.

Another technique that I'm just learning about is unbelievably slippery; it is called an INTERSPERSAL TECHNIQUE and the idea is to say one thing with words but plant a subconscious impression of something else in the minds of the listeners and/or watchers.

Let me give you an example: Assume you are watching a television commentator make the following statement: "SENATOR JOHNSON is assisting local authorities to clear up the stupid mistakes of companies contributing to the nuclear waste problems." It sounds like a statement of fact, but, if the speaker emphasizes the right word, and especially if he makes the proper hand gestures on the key words, you could be left with the subconscious impression that Senator Johnson is stupid. That was the subliminal goal of the statement and the speaker cannot be called to account for anything.

Persuasion techniques are also frequently used on a much smaller scale with just as much effectiveness. The insurance salesman knows his pitch is likely to be much more effective if he can get you to visualize something in your mind. This is right-brain communication. For instance, he might pause in his conversation, look slowly around your living room and say, "Can you just imagine this beautiful home burning to the ground?" Of course, you can! It is one of your unconscious fears and, when he forces you to visualize it; you are more likely to be manipulated into signing his insurance policy.

The Hare Krishnas, operating in every airport, use what I call SHOCK AND CONFUSION techniques to distract the left brain and communicate directly with the right brain. While waiting for a plane, I once watched one operate for over an hour. He had a technique of almost jumping in front of someone. Initially, his voice was loud then dropped as he made his pitch to take a book and contribute money to the cause. Usually, when people are shocked, they immediately withdraw. In this case they were shocked by the strange appearance, sudden materialization and loud voice of the Hare Krishna devotee. In other words, the people went into an alpha state for security because they didn't want to confront the reality before them. In alpha, they were highly suggestible so they responded to the suggestion of taking the book; the moment they took the book, they felt guilty and responded to the second suggestion: give money. We are all conditioned that if someone gives us something, we have to give them something in return; in that case, it was money. While watching this hustler, I was close enough to notice that many of the people he stopped exhibited an outward sign of alpha, their eyes were actually dilated.

Subliminal Programming

Subliminals are hidden suggestions that only your subconscious perceives. They can be audio, hidden behind music, or visual, airbrushed into a picture, flashed on a screen so fast that you don't consciously see them, or cleverly incorporated into a picture or design.

Most audio subliminal reprogramming tapes offer verbal suggestions recorded at a low volume. I question the efficacy of this technique if subliminals are not perceptible, they cannot be effective, and subliminals recorded below the audible threshold are therefore useless. The oldest audio subliminal technique uses a voice that follows the volume of the music so subliminals are impossible to detect without a parametric equalizer. But this technique is patented and, when I wanted to develop my own line of subliminal audiocassettes, negotiations with the patent holder proved to be unsatisfactory. My attorney obtained copies of the patents which I gave to some talented Hollywood sound engineers, asking them to create a new technique. They found a way to psycho-acoustically modify and synthesize the suggestions so that they are projected in the same chord and frequency as the music, thus giving them the effect of being part of the music. But we found that in using this technique, there is no way to reduce various frequencies to detect the subliminals. In other words, although the suggestions are being heard by the subconscious mind, they cannot be monitored with even the most sophisticated equipment.

If we were able to come up with this technique as easily as we did, I can only imagine how sophisticated the technology has become, with unlimited government or advertising funding. And I shudder to think about the propaganda and commercial manipulation that we are exposed to on a daily basis. There is simply no way to know what is behind the music you hear. It may even be possible to hide a second voice behind the voice to which you are listening. The series by Wilson Bryan Key, Ph.D., on subliminals in advertising and political campaigns well documents the misuse in many areas, especially printed advertising in newspapers, magazines, and posters.

The big question about subliminals is: do they work? And I guarantee you they do. Not only from the response of those who have used my tapes, but from the results of such programs as the subliminals behind the music in department stores. Supposedly, the only message is instructions to not steal: one East Coast department store chain reported a 37 percent reduction in thefts in the first nine months of testing.

A 1984 article in the technical newsletter, "Brain-Mind Bulletin," states that as much as 99 percent of our cognitive activity may be "non-conscious," according to the director of the Laboratory for Cognitive Psychophysiology at the University of Illinois. The lengthy report ends with the statement, "these findings support the use of subliminal approaches such as taped suggestions for weight loss and the therapeutic use of hypnosis and Neuro-Linguistic Programming."

Mass Misuse

I could relate many stories that support subliminal programming, but I'd rather use my time to make you aware of even more subtle uses of such programming.

I have personally experienced sitting in a Los Angeles auditorium with over ten thousand people who were gathered to listen to a current charismatic figure. Twenty minutes after entering the auditorium, I became aware that I was going in and out of an altered state. Those accompanying me experienced the same thing. Since it is our business, we were aware of what was happening, but those around us were not.

The only way I could figure that the eyes-open trance had been induced was that a 6- to 7-cycle-per-second vibration was being piped into the room behind the air conditioner sound. That particular vibration generates alpha, which would render the audience highly susceptible. Ten to 25 percent of the population is capable of a somnambulistic level of altered states of consciousness; for these people, the suggestions of the speaker, if non-threatening, could potentially be accepted as "commands."

Vibrato

This leads to the mention of VIBRATO. Vibrato is the tremulous effect imparted in some vocal or instrumental music, and the cycle-per-second range causes people to go into an altered state of consciousness. At one period of English history, singers whose voices contained pronounced vibrato were not allowed to perform publicly because listeners would go into an altered state and have fantasies, often sexual in nature.

People who attend opera or enjoy listening to singers like Mario Lanza are familiar with this altered state induced by the performers.

Extra Low Frequencies

Now, let's carry this awareness a little farther. There are also inaudible ELFs (extra-low frequency waves). These are electromagnetic in nature. One of the primary uses of ELFs is to communicate with our submarines. Dr. Andrija Puharich, a highly respected researcher, in an attempt to warn U.S. officials about Russian use of ELFs, set up an experiment. Volunteers were wired so their brain waves could be measured on an EEG. They were sealed in a metal room that could not be penetrated by a normal signal.

Puharich then beamed ELF waves at the volunteers. ELFs go right through the earth and, of course, right through metal walls. Those inside couldn't know if the signal was or was not being sent. And Puharich watched the reactions on the technical equipment: 30 percent of those inside the room were taken over by the ELF signal in six to ten seconds.

When I say "taken over," I mean that their behavior followed the changes anticipated at very precise frequencies. Waves below 6 cycles per second caused the subjects to become very emotionally upset and even disrupted bodily functions. At 8.2 cycles, they felt very high… an elevated feeling, as though they had been in masterful meditation, learned over a period of years. Eleven to 11.3 cycles induced waves of depressed agitation leading to riotous behavior.

The Neurophone

Dr. Patrick Flanagan is a personal friend of mine. In the early 1960s, as a teenager, Pat was listed as one of the top scientists in the world by "Life" magazine. Among his many inventions was a device he called the Neurophone an electronic instrument that can successfully program suggestions directly through contact with the skin. When he attempted to patent the device, the government demanded that he prove it worked. When he did, the National Security Agency confiscated the Neurophone. It took Pat two years of legal battle to get his invention back.

In using the device, you don't hear or see a thing; it is applied to the skin, which Pat claims is the source of special senses. The skin contains more sensors for heat, touch, pain, vibration, and electrical fields than any other part of the human anatomy.

In one of his recent tests, Pat conducted two identical seminars for a military audience one seminar one night and one the next night, because the size of the room was not large enough to accommodate all of them at one time. When the first group proved to be very cool and unwilling to respond, Patrick spent the next day making a special tape to play at the second seminar. The tape instructed the audience to be extremely warm and responsive and for their hands to become "tingly." The tape was played through the Neurophone, which was connected to a wire he placed along the ceiling of the room. There were no speakers, so no sound could be heard, yet the message was successfully transmitted from that wire directly into the brains of the audience. They were warm and receptive, their hands tingled and they responded, according to programming, in other ways that I cannot mention here.

The more we find out about how human beings work through today's highly advanced technological research, the more we learn to control human beings. And what probably scares me the most is that the medium for takeover is already in place! The television set in your living room and bedroom is doing a lot more than just entertaining you.

Before I continue, let me point out something else about an altered state of consciousness. When you go into an altered state, you transfer into right brain, which results in the internal release of the body's own opiates: enkephalins and Beta-endorphins, chemically almost identical to opium. In other words, it feels good and you want to come back for more.

Recent tests by researcher Herbert Krugman showed that, while viewers were watching TV, right-brain activity outnumbered left-brain activity by a ratio of two to one. Put more simply, the viewers were in an altered state . . . in trance more often than not. They were getting their Beta-endorphin "fix."

To measure attention spans, psycho-physiologist Thomas Mulholland of the Veterans Hospital in Bedford, Massachusetts, attached young viewers to an EEG machine that was wired to shut the TV set off whenever the children's brains produced a majority of alpha waves. Although the children were told to concentrate, only a few could keep the set on for more than 30 seconds!

Most viewers are already hypnotized. To deepen the trance is easy. One simple way is to place every 32 frames in the film that is being projected. This creates a 45-beat-per-minute pulsation perceived only by the subconscious mind the ideal pace to generate deep hypnosis.

The commercials or suggestions presented following this alpha-inducing broadcast are much more likely to be accepted by the viewer. The high percentage of the viewing audience that has somnambulistic-depth ability could very well accept the suggestions as commands as long as those commands did not ask the viewer to do something contrary to his morals, religion, or self-preservation.

The medium for takeover is here. By the age of 16, children have spent 10,000 to 15,000 hours watching television that is more time than they spend in school! In the average home, the TV set is on for six hours and 44 minutes per Dayan increase of nine minutes from last year and three times the average rate of increase during the 1970s.

It obviously isn't getting better . . . we are rapidly moving into an alpha-level world, very possibly the Orwellian world of "1984"placid, glassy-eyed, and responding obediently to instructions.

A research project by Jacob Jacoby, a Purdue University psychologist, found that of 2,700 people tested, 90 percent misunderstood even such simple viewing fare as commercials and "Barnaby Jones." Only minutes after watching, the typical viewer missed 23 to 36 percent of the questions about what he or she had seen. Of course, they did, they were going in and out of trance! If you go into a deep trance, you must be instructed to remember, otherwise you automatically forget.

I have just touched the tip of the iceberg. When you start to combine subliminal messages behind the music, subliminal visuals projected on the screen, hypnotically produced visual effects, sustained musical beats at a trance-inducing pace . . . you have extremely effective brainwashing. Every hour that you spend watching the TV set you become more conditioned. And, in case you thought there was a law against any of these things, guess again. There isn't! There are a lot of powerful people who obviously prefer things exactly the way they are. Maybe they have plans for us?

This article was reproduced in Fact, Fiction and Fraud in Modern Medicine in February 1999.

NEUROPHONE: U.S. Patent #3,393,279.

July 16, 1968. Inventor - Dr. Patrick Flanagan (Invented in 1958)

DESCRIPTION: A device that converts sound to electrical impulses; allowing information to be transmitted to the brain by means of radio waves directed at any part of the body (skin). In other words, recorded or live messages, noise, music can be directed at an individual and, through the nerves, the signal will be carried (involuntarily) to the brain, bypassing the inner ear, the cochlea, and the 8th cranial nerve. In its original form electrodes were placed on the skin but with defense department developments, the signals can be delivered via satellite.

PURPOSE: Practically, the Neurophone could be used to communicate with the deaf but more often, it is used to terrorize targets. The tracked individual's here recorded/live threats, propaganda, etc. which those around them do not hear (delivered mainly via satellite laser). This harasses and discredit's the targets; especially if the problem is communicated to those unaware or the relevant technologies.

ADVANCED NEUROPHONE: U. S. Patent # 3,647,970

March 7, 1972. Inventor - Dr. Patrick Flanagan (Invented 1967)

DESCRIPTON: This Neurophone incorporates an electronic circuit duplicating the encoding of the Cochlea and 8th cranial nerve themselves. The NSA placed a secrecy order on this development for over 5 years because of the military applications of the technology.

Further Neurophone advances include the development of the time recognition processor, improved memory application and the advances in satellites incorporating neurophone technologies.

PURPOSE: As above.

CHAPTER THREE

"An illusion can become a half-truth; a mask can alter the expression of a face. The familiar arguments to the effect that democracy is 'just the same as' or 'just as bad as' totalitarianism never take account of this fact."

- George Orwell (1984)

"The experience of Artificial Telepathy or Synthetic Telepathy is really not that extraordinary" says the administrator of the mindtechsweden.com website. "It's as simple as receiving a cell-phone call in one's head. Indeed, the site administrator states, most of the technology involved is exactly identical to that of cell-phone technology. Satellites link the sender and the receiver. A computer "multiplexer" routes the voice signal of the sender through microwave towers to a very specifically defined location or cell. The "receiver" is located and tracked with pinpoint accuracy, to within a few feet of actual location. But the receiver is not a cell phone. It's a human brain.

Out of nowhere, a voice suddenly blooms in the mind of the target. The human skull has no "firewall" and therefore cannot shut the voice out. The receiver can hear the sender's verbal thoughts. The sender, in turn, can hear all of the target's thoughts exactly as if the target's verbal thoughts had been spoken or broadcast. For this reason, the experience could be called "hearing voices" but is more properly described as "artificial telepathy". Other names for the capability to communicate directly into the human brain from miles away are: Voice to Skull (V2K), Neural Decoding, Neurophone, and the Frey Effect.

The very first Blog Talk Radio internet talk show interview I did in late 2012 was with Dr. Richard Alan Miller, bestselling author, physicist and retired Navy intelligence officer on "The Targeted Admittedly, I was an amateur who fell into the broadcast under duress through energy weapons attacks and efforts to sabotage me.

With Dr. Miller's permission, I included his work's main focus, due to his thorough presentation:

SYNTHETIC TELEPATHY AND THE EARLY MIND WARS

By Richard Alan Miller

Presented at the Consciousness Technologies Conference

July 19-21, 2001, Saturday, July 20, 2001, in Sisters, OR

Updated 03/04/2003

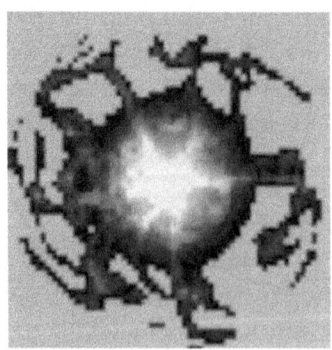

Introduction

I first became aware of Allan H. Frey's work at Willow Grove in 1972, just after completing "The Holographic Concept of Reality." I was working with Dr. Carl Scheicher (MRU) at the time, and was asked about the significance of this work. Realizing its possible use in mind control, my first reaction was to go on "red alert."

Full significance was not yet understood at this time. Basically, Frey had discovered another sensory motor input in the higher blue-band frequencies of 0.3-3.0 GHz. - at very low amplitudes of power. It was "as if" we had another type of "vision," but did not know how to "see" what was being received. It constituted the next generation of subliminal communications.

My work at the time was involved with an AI database for paranormal references (Project Parafile). A second paper was also presented at the Omniversal Symposium, California State College at Sonoma, (September 29, 1973). This was titled "Embryonic Holography," and was an application of "The Holographic Concept of Reality" model. It dealt with biogenesis and neurological regeneration, and included speculation on the origins of cancer, faith healing, psychic surgery and more technical aspects of mind-body energetics.

One week after the delivery of that paper, four men came into my place of business, two in suits and two in full Army dress. The two suits held me under close arrest, while the two Army personnel went through my files, pulling anything related to "Embryonic Holography." The paper was rewritten from old notes and memory, but it was not the same.

What got this paper classified "top secret" for almost 20 years was that it was critical for the use of Allan Frey's study, and its possible application to mind control. I never was able to draw what was so important in that initial paper until I began researching this paper, more than 24 years later. I will discuss those aspects further into this document.

In 1961, Allan Frey, a freelance biophysicist and engineering psychologist, reported that humans could hear microwaves. Most United States scientists dismissed this discovery as the result of outside noise.

James C. Linn offered a more technical description of the experiment.

"Frey found that human subjects exposed to 1310 MHz and 2982 MHz microwaves at average power densities of 0.4 to 2 mW/cm2 perceived auditory sensations described as buzzing or knocking sounds. The sensation occurred instantaneously at average incident power densities well below that necessary for known biological damage and appeared to originate from within or near the back of the head."

Pulsed Microwave Technology

Pulsed microwave voice-to-skull (or other-sound-to-skull) transmission was discovered during World War II by radar technicians who found they could hear the buzz of the train of pulses being transmitted by radar equipment they were working on. This phenomenon has been studied extensively by Dr. Allan Frey, (Willow Grove, 1965) whose work has been published in a number of reference books.

What Dr. Frey found was that single pulses of microwave could be heard by some people as "pops" or "clicks", while a train of uniform pulses could be heard as a buzz, without benefit of any type of receiver.

Dr. Frey also found that a wide range of frequencies, as low as 125 MHz (well below microwave) worked for some combination of pulse power and pulse width. Detailed unclassified studies mapped out those frequencies and pulse characteristics which are optimum for generation of "microwave hearing".

Very significantly, when discussing electronic mind control, is the fact that the peak pulse power required is modest - something like 0.3 watts per square centimeter of skull surface, and that this power level is only applied or needed for a very small percentage of each pulse's cycle time. 0.3-watts/sq cm is about what you get under a 250-watt heat lamp at a distance of one meter. It is not a lot of power.

When you take into account that the pulse train is off (no signal) for most of each cycle, the average power is so low as to be nearly undetectable. This is the concept of "spike" waves used in radar and other military forms of communication.

Frequencies that act as voice-to-skull carriers are not single frequencies, as, for example TV or cell phone channels. Each sensitive frequency is actually a range or "band" of frequencies. A technology used to reduce both interference and detection is called "spread spectrum". Spread spectrum signals usually have the carrier frequency "hop" around within a specified band.

Unless a receiver "knows" this hop schedule in advance, like other forms of encryption there is virtually no chance of receiving or detecting a coherent readable signal. Spectrum analyzers, used for detection, are receivers with a screen. A spread spectrum signal received on a spectrum analyzer appears as just more "static" or noise.

The actual method of the first successful unclassified voice to skull experiment was in 1974, by Dr. Joseph C. Sharp and Mark Grove, then at the Walter Reed Army Institute of Research. A Frey-type audible pulse was transmitted every time the voice waveform passed down through the zero axes, a technique easily duplicated by ham radio operators who build their own equipment.

The sensation is reported as a buzzing, clicking, or hissing which seems to originate within or just behind the head. The phenomenon occurs with carrier densities as low as microwatts per square centimeter with carrier frequencies from 0.3-3.0 GHz. By proper choice of pulse characteristics, intelligent speech may be created.

Dr. James Lin of Wayne State University has written a book entitled: Microwave Auditory Effects and Applications. It explores the possible mechanisms for the phenomenon, and discusses possibilities for the deaf, as persons with certain types of hearing loss can still hear pulsed microwaves (as tones or clicks and buzzes, if words aren't modulated on). Lin mentions the Sharp and Grove experiment and comments: "The capability of communicating directly with humans by pulsed microwaves is obviously not limited to the field of therapeutic medicine."

"Synthetic Telepathy"

In 1975, researcher A. W. Guy stated that "one of the most widely observed and accepted biologic effects of low average power electromagnetic energy is the auditory sensation evoked in man when exposed to pulsed microwaves."

He concluded that at frequencies where the auditory effect can be easily detected, microwaves penetrate deep into the tissues of the head, causing rapid thermal expansion (at the microscopic level only) that produces strains in the brain tissue.

An acoustic stress wave is then conducted through the skull to the cochlea, and from there, it proceeds in the same manner as in conventional hearing. It is obvious that receiver-less radio has not been adequately publicized or explained because of national security concerns.

Today, the ability to remotely transmit microwave voices inside a target's head is known inside the Pentagon as "Synthetic Telepathy." According to Dr. Robert Becker,

"Synthetic Telepathy has applications in covert operations designed to drive a target crazy with voices or deliver undetected instructions to a programmed assassin."

This technology may have contributed to the deaths of 25 defense scientists variously employed by Marconi Underwater and Defense Systems, Easems and GEC. Most of the scientists worked on highly sensitive electronic warfare programs for NATO, including the Strategic Defense Initiative. It is claimed that directed energy weapons might have been used to literally drive these men to suicide and 291accidents.

Biological Amplification Using Microwave Band Frequencies

The next major development in ELF weaponry was the concept of a biological amplification of these signals at the cell level to perpetuate and set up resonance for more sophisticated information transfer. This was the beginning of using more than one technology in a stack to do something "more." While this was implied, it was never developed in "The Holographic Concept of Reality."

Electromagnetic fields or relatively weak power levels can affect intercellular communication. Bio-amplification is apparently why radio signals of very low average power (mw) can produce audio effects, and is difficult to detect. [Electromagnetic Interaction with Biological Systems, ed. Dr. James C. Lin, Univ. of Illinois, 1989, Plenum Press, NY]

Imposed weak low frequency fields (and radio frequency fields) that are many orders of magnitude weaker in the pericellular fluid (fluid between adjacent cells) than the membrane potential gradient (voltage across the membrane) can modulate the action of hormones, antibody neurotransmitters and cancer-promoting molecules at their cell surface receptor sites.

These ELF sensitivities appear to involve nonequilibrium and highly cooperative processes that mediate a major amplification of initial weak triggers associated with the binding of these molecules (specific cell surface receptor sites). Membrane amplification is inherent in this trans-membrane signaling sequence.

Initial stimuli associated with weak perpendicular EM fields and with binding of stimulating molecules at their membrane receptor sites elicit a highly cooperative modification of Ca++ binding to glycoproteins along the membrane.

A longitudinal spread is consistent with the direction of extracellular current flow associated with physiological activity and imposed EM fields. This cooperative modification of surface Ca++ binding is an amplifying stage. By imposing RF fields, there is a far greater increase in Ca++ efflux than is accounted for in the events of receptor-ligand binding. from imposing RF fields.

Enzymes are protein molecules that function as catalysts, initiating and enhancing chemical reactions that would not otherwise occur at tissue temperatures. This ability resides in the pattern of electrical charges on the molecular surface.

Activation of these enzymes and the reactions in which they participate involve energies millions of times greater than in the cell surface, triggering events initiated by the EM fields, emphasizing the membrane amplification inherent in this trans-membrane signaling sequence.

Frey and Messenger confirmed that a microwave pulse with a slow rise time was ineffectual in producing an auditory response. Only if the rise time is short, resulting in effect in a square wave with respect to the leading edge of the envelope of radiated radio-frequency energy, does the auditory response occur. This is why we don't "hear" ordinary radio and TV signals.

The significance of "Embryonic Holography" now becomes more understandable. For example, the specific frequency bands (0.3-3.0 Hz) are so flat as to appear almost 2-dimensional to most biological processes on a semi-quantum mechanical level. This means that these frequencies can be seen as "scalar" in their possible interaction with specific brain processes.

What these frequencies really are, however, are actual holograms of specific thoughts. They have a third component of detail (much like the patented P300 wave). This means that a hybrid form of brain fingerprinting is now possible. And, once these "images" are stored (usually in a very sophisticated super-cooled computer), similar responses can be fed back to the person, inducing virtually any state desired (via entrainment protocols).

Silent Sound Technology - "S-quad"

Silent (converted-to-voice FM) hypnosis can be transmitted using a voice frequency modulator to generate the "voice." It is a steady tone, near the high end of hearing range (15,000 Hz), plus a hypnotist's voice, varying from 300 - 4,000 Hz. These two signals are frequency modulated. The output now appears as a steady tone, like tinnitus, but with hypnosis embedded. The FM-voice controls the timing of the transmitter's pulse.

Each vertical line is one short pulse of microwave signal at a frequency to which the human brain is sensitive. Timing of each microwave pulse is controlled by each down-slope crossing of the voice wave (Sharp's method, 1974). Then the brain converts the train of microwave pulses back to inaudible voice. There is no conscious defense possible against this form of hypnosis.

Ordinary radio and TV signals use a smooth waveform called a 'sine' wave. This wave signal cannot normally penetrate the voltage gradient across the nerve cell walls. Radar signals consist of very short and powerful pulses of sine wave type signals, and can penetrate the steep voltage gradient across these nerve cell walls (Allan H. Frey, Cornell University, 1962).

Differences in osmosis of ions (dissolved salt components) cause a small voltage difference across cell walls. When a small voltage appears across a very tiny distance, the change in voltage is called very 'steep.' It is this steep gradient that keeps normal radio signals from throwing us into convulsions.

The mind-altering mechanism is based on a subliminal carrier technology: The Silent Sound Spread Spectrum (SSSS) sometimes called "S-quad" or "Squad". It was developed by Dr. Oliver Lowery of Norcross, Georgia, and is described in US Patent #5,159,703, "Silent Subliminal Presentation System", dated October 27, 1992. The abstract for the patent reads:

"A silent communications system in which non-aural carriers, in the very low or very high audio-frequency range or in the adjacent ultrasonic frequency spectrum are amplitude- or frequency-modulated with the desired intelligence and propagated acoustically or vibrationally, for inducement into the brain, typically through the use of loudspeakers, earphones, or piezoelectric transducers. The modulated carriers may be transmitted directly in real time or may be conveniently recorded and stored on mechanical, magnetic, or optical media for delayed or repeated transmission to the listener."

According to literature by Silent Sounds, Inc., it is now possible, using supercomputers, to analyze human emotional EEG patterns and replicate them, then store these "emotion signature clusters" on another computer and, at will, "silently induce and change the emotional state in a human being".

Edward Tilton, President of Silent Sounds, Inc. says this about S-quad in a letter dated December 13, 1996:

"All schematics, however, have been classified by the US Government and we are not allowed to reveal the exact details... ... we make tapes and CDs for the German Government, even the former Soviet Union countries! All with the permission of the US State Department, of course... The system was used throughout Operation Desert Storm (Iraq) quite successfully."

"Induced Alpha to Theta Biofeedback Cluster Movement" is an output from "the world's most versatile and most sensitive electroencephalograph (EEG) machine". This device has a gain capability of 200,000, as compared to most other EEG machines (with gain capability of 50,000). It is software-driven by the "fastest of computers" using a noise nulling technology similar to that used by nuclear submarines for detecting small objects underwater at extreme range.

The purpose of all this high technology is to plot and display a moving cluster of periodic brainwave signals. The illustration shows an EEG display from a single individual, taken of left and right hemispheres simultaneously. This technology is very similar to that used to generate P300 waves.

Cloning the Emotions

By using these computer-enhanced EEGs, scientists can identify and isolate the brain's low-amplitude "emotion signature clusters," synthesize them and store them on another computer. In other words, by studying the subtle characteristic brainwave patterns that occur when a subject experiences a particular emotion, scientists have been able to identify the concomitant brainwave pattern and can now duplicate it.

"These clusters are then placed on the Silent Sound[TM] carrier frequencies and will silently trigger the occurrence of the same basic emotion in another human being!"

Regarding system delivery and applications, there is a lot more involved here than a simple subliminal sound system. There are numerous patented technologies that can be piggybacked individually or collectively onto a carrier frequency to elicit all kinds of effects.

There appear to be two methods of delivery with the system. One is direct microwave induction into the brain of the subject, limited to short-range operations. The other, as described above, utilizes ordinary radio and television carrier frequencies.

Far from necessarily being used as a weapon against a person, the system does have limitless positive applications. However, the fact that the sounds are subliminal makes them virtually undetectable and possibly dangerous to the general public.

In more conventional use, the Silent Sounds Subliminal System might utilize voice commands, e.g., as an adjunct to security systems. Beneath the musical broadcast that you hear in stores and shopping malls may be a hidden message that exhorts against shoplifting. And while voice commands alone are powerful, when the subliminal presentation system carries cloned emotional signatures, the result is overwhelming.

Free-market uses for this technology are the common self-help tapes, positive affirmation, relaxation and meditation tapes, as well as methods to increase learning capabilities. But there is strong evidence that this technology is being developed toward global mind control.

The secrecy involved in the development of the electromagnetic mind-altering technology reflects the tremendous power that is inherent in it. To put it bluntly, whoever controls this technology can control the minds of men - all men.

There is evidence that the U.S. Government has plans to extend the range of this technology to envelop all peoples, all countries. This can be accomplished, and is being accomplished, by utilizing the nearly completed HAARP project for overseas areas and the GWEN network now in place in the U.S. The U.S. Government denies all this.

Dr. Michael Persinger is a Professor of Psychology and Neuroscience at Laurentian University, Ontario, Canada. His work and findings indicate that strong electromagnetic fields can and will affect a person's brain.

"Temporal lobe stimulation can evoke the feeling of a presence, disorientation, and perceptual irregularities. It can activate images stored in the subject's memory, including nightmares and monsters that are normally suppressed."

Mind Reading Devices

Alan Yu, a former lieutenant colonel in the Taiwan National Defense Department, says that the United States has not only developed an operational mind control machine, but has also distributed models for use by allied countries. Yu states that such machines pose a great threat to human rights and the American way of life. He calls the device the "Mind Reading Machine" (Mind Machine).

Yu writes that there are two sources of information detailing the existence of the Mind Reading Machine. The first evidence: In the 1970s, The South China Morning Post reported that the University of Maryland had invented a Thought Reading Machine. The original purpose of this invention was to help authorities investigate severe car accidents. It was to be used on people who were severely injured to get their accounts of how the accidents occurred.

In the spring of 1984, Yu was a lieutenant colonel serving in the National Defense Department of Taiwan. At that time, Yu read a classified document from the department that he serviced under. The document said the Military Police Department of Taiwan had purchased several of the Mind Reading Machines from the United States (In Taiwan, it was called Psychological Language Machine).

The document was a request to the United States for parts to repair several malfunctioning machines. The machine allegedly uses microwaves to deliver spoken messages directly to the human brain, as well as using radio waves to hypnotize people or change their thoughts. Yu reports that before he left, this machine had become the most effective weapon for the security departments of Taiwan.

In 1993, Defense News announced that the Russian government was discussing with American counterparts the transfer of technical information and equipment known as "Acoustic Psycho-correction." The Russians claimed that this device involves the transmission of specific commands via static or white noise bands into the human subconscious without upsetting other intellectual functions.

Demonstrations of this equipment have shown encouraging results after exposure of less than one minute and have produced the ability to alter behavior on unwilling subjects. A U.S. Department of Defense medical engineer claimed in 1989 that the U.S. and Israel had regularly used microwaves to condition and control the minds of Palestinians.

Unclassified ELF-Type Weapons

Remote physical manipulation: Not covered in this document. At time of writing, that technology appears to be classified.

Transmission methods for neuro-effective signals: This includes pulsed microwave (i.e., like radar signals) and ultrasound and voice-FM (transmitted through the air), which is also known as "Synthetic Telepathy."

While transmission of speech, dating from the early 1970s, was the first use of pulsed microwave, neuro-effective signals can now cause many other nerve groups to become remotely actuated. That specific technology is classified.

Pavlovian hypnotic triggers: A [Pavlovian] hypnotic trigger is a phrase or any sensory cue that the subject is programmed to involuntarily act in a certain way. MKULTRA survivors can still be triggered from programming done decades ago.

One of the main goals of the institutional/drug/child abuse phases of the CIA MKULTRA atrocities (1950's through 1970's) was to implant triggers using a "twilight state" (half-conscious) medication and tape-recorded hypnosis. The ultimate goal was to have the acting out of Pavlovian triggers erased from the victim's memory.

These triggers are now planted using either of the above two transmission methods, but with the words moved up just above (or near the top of) the audible frequency range. The result is that hypnotic triggers are planted without the subject being aware. This technology was used in the Gulf War as "Silent Sound."

Through-wall surveillance methods: This includes top end of microwave (near infrared), and the so-called "millimeter wave" scanning. This method uses the very top end of the microwave radio signal spectrum just below infrared. To view small objects or people clearly, the highest frequency that will penetrate non-conductive or poorly-conductive walls is used.

Millimeter wave scanning radar can be used in two modes: The first is passive (no radiated signal) and uses background radiation already in the area to be scanned. It is totally undetectable. The active system uses a (low power millimeter wave) "flashlight" attached to the scanner.

Thought reading: Thought reading can be classed as a form of "through wall surveillance" technology. In the unclassified and commercial realms, it is called thru-skull microwave reading, and magnetic skull-proximity reading.

Brain entrainment: This involves moods and sleep states, the reverse of biofeedback. The low frequency electrical brain rhythms are characteristics of various moods and states of sleep. Not only can they now be read out using biofeedback equipment or EEG machines, but also radio, sound, contact electrodes, or flashing lights. These moods and sleep states can be generated or at least encouraged using brain entrainment devices.

Brain entrainment signals cannot carry voice, which is a much higher frequency range. Brain entrainment can, however, be used to "set up" a target to make him/her more susceptible to hypnosis.

Implantation: (no longer required)

Specific ELF Weapons

Ultrasound and Voice-FM: Main advantage in mind control work is that it can carry verbal hypnosis, more potent than simple biorhythm entrainment. An example is Chicago's Airport Terminal connection tunnels and their "Keep Walking."

Steady tone, near the high end of hearing range (15,000 Hz). Hypnotist's voice, varying from 300 to 4,000 Hz, fed into a frequency modulator, where the voice controls the frequency. Output is now a steady tone, sounding like tinnitus, but with hypnosis embedded. While the brain can hear and understand, the ear only hears a "tone" or a "rush."

Acoustic Heterodyne
American Technologies Corp.
13114 Evening Creek Drive. S.
San Diego, CA 92128

Through-Wall Radar: Millimeter wave through clothing, through-luggage is currently in use at airports. Millimeter wave scanners can be purchased from:

Millivision Corp.
Northampton, MA
www.millivision.com

Thought Readings: Thought reading is an enhanced version of computer speech recognition, with EEG waves being substituted for sound waves. The easiest "thought" reading is actually remote picking up of the electro-magnetic activity of the speech-control muscles.

When we say words to ourselves, silently, or, read a book, we can actually feel the slight sensations of those words in our vocal muscles - all that is absent is the passage of air. Coordinated speech signals are relatively strong and relatively consistent.

We are "fed" hypnotic signals to force consistent "neutral" content (but of different character than prior to becoming test subjects) in dreams. These forced, neutral content ("bland" content) dreams occur every single night and may represent the experimenters' efforts to have our experiences portray themselves in such dreams, in effect, mining our experiences.

www.raven1.net/elecvisn.htm confirms the ability of current unclassified technology to actually see what a living animal sees, electronically. It is therefore extremely likely that these forced dreams can be displayed on the experimenters' screens in an adjacent apartment or adjacent house, (which are made obvious to the involuntary experimentee).

Implants: Implants can either receive instructions via radio signals, passing them to the brain. Or, can be interrogated via external radio signals to read brain activity at a distance. Since implants for beneficial purposes are actively being promoted by NIH, it is obvious they will not disappear any time soon.

Thermal Gun, Seizure Gun, and Magnetophosphene Gun: Evokes a visual response and is thought to be centered in the retina (as seen in the movies Goldeneye, Broken Arrow, Escape From LA, and Eraser). The popular video "Waco: The Big Lie Continues" shows video footage of three EM weapons being used during the confrontation.

Silent Subliminals

ALTERED STATES LTD
P.O.Box 68-344, Newton,
Auckland, New Zealand.
Ph: +64-9-815-5095 or +64-9-815-5059
Fax: +64-9-815-5067
altered@ihug.co.nz
//www.altered-state.com/index2.htm

Note the format for this website has changed; please refer to this information under "Understanding Neurosync"

From their brochure:

"Only your mind can hear. Your ears hear nothing but your mind hears and accepts the powerful suggestions.

You can safely play these tapes anywhere - in a car, while watching TV or listening to your favorite music, while working or even as silent sleep programming."

Warning: Everyone within listening range of the tape will be programmed by the suggestions. To assure yourself that strong suggestions are recorded on the tape, take it to any Radio Shack store, play it on their stereo and read the output with a Radio Shack Sound Level Meter (Item 33-2050)

How to Use the Tapes: Increase the volume until it is just below any tape noise. If your stereo deck has treble and bass controls, you can boost the subliminal output by increasing the treble and decreasing the bass. The player then emits a strong but inaudible frequency - modulated 60 - 90 decibel signal that is received and demodulated by the human ear.

Technical Information: The Suggestions are delivered on a carrier frequency of 14,800 cps, via a low-distortion sine wave signal. This frequency is slightly above the audible hearing range but the frequency-modulated (FM) signal is still strongly impinging upon the diaphragm of the ear. The listener can expect his subconscious mind to accept the suggestions with repeated listening.

The Silent Subliminals is a new brain/mind technology developed by an aerospace engineer. This new technique has been licensed to Valley of the Sun Audio/Video for this incredible new tape series. Patent pending. Note: Because the frequency is beyond normal recording range, the tape cannot be duplicated:

Examples of Suggestions:

- "Every day you become thinner and thinner"

- "You now lose weight and full fill your goals"

- "You attain your weight goals and the body you desire"

- "You have the power and ability to attain the perfect weight

- and body you desire"
- "You have the self-discipline to lose all the weight
- you want"
- "You live a healthy lifestyle and eat a proper diet"
- "You now quit smoking because it serves you"
- "You lose all desire to smoke"
- "You accept that you now quit smoking"
- "You are a non-smoker"
- "Quit smoking. Quit smoking. Quit smoking"
- "You have the willpower to do anything you want to do"
- "You have great self- discipline and you use it
- to quit smoking"
- "Cigarettes disgust you"
- "You are very sure of yourself"
- "You accept that you have great inner courage"
- "You are self-reliant and self-confident"
- "You are full of independence and determination"
- "You have great inner courage"
- "Every day in every way, you become more and more self-confident"
- "You feel good about yourself"

- "You project a very positive self-image"
- "You are relaxed and at ease"
- "You detach from worldly pressures and experience an inner calm"
- "Negativity flows through you without affecting you"
- "You accept other people as they are"
- "You peacefully accept the things you cannot change, and change the things you can"
- "You are at peace with yourself, the world and everyone in it"
- "Your mind is like calm water
- "You direct your time and energy to manifest your desires"
- "You have the self-discipline to accomplish your personal and professional goals"
- "Every day, you increase your self-discipline"
- "You do what you need to do and stop doing what doesn't work"
- "You freely choose to do what you need to do"
- "You are assertive and feel good about yourself
- "You now focus your energy upon attaining success"
- "You know exactly what you want and you go for it"
- "You can accomplish whatever you set out to do"

- "Be ultra-successful. Be ultra-successful and become wealthy"

- "Every day in every way, you become more successful"

- "Your creative thinking opens the door to monetary abundance"

- "You easily achieve and maintain a penile erection"

- "Your body performs perfectly during sex without thinking about it"

- "A hard, firm erection is your natural response to sexual stimulation "

- "You can make love for a long before you ejaculate"

- "Every day you feel better about your sexual prowess and your ability to achieve and maintain a hard, firm erection"

"Acoustic Spotlight (Can target one person in crowd)"

Posted by F. Joseph Pompeii, MIT Media Lab

Usage: The Audio Spotlight can be used in two major ways: As directed audio, sound is directed at a specific listener or area, to provide a private or area specific listening space. As projected audio, sound is projected against a distant object, creating an audio image. This audio image is literally a projected loudspeaker - sound appears to come directly from the projection, just like light.

The Audio Spotlight consists of a thin, circular transducer array and a specially designed signal processor and amplifier. The transducer is about half an inch thick, nonmagnetic, and lightweight. The signal processor and amplifier are integrated into a unit about the same size as a traditional audio amplifier, and have similar power requirements.

Technology: Because it is impossible to generate extremely narrow beams of audible sound without extremely large loudspeaker arrays, we instead generate the sound indirectly, using the nonlinearity of the air to convert a narrow beam of ultrasound into a highly directive, audible beam of sound.

The device transmits a narrow beam of ultrasound (blue), which, due to the inherent nonlinearity of the air itself, distorts (changes shape) very slightly as it travels. This distortion creates, along with new ultrasonic frequencies, audible artifacts (green) that can be mathematically predicted, and therefore controlled.

By constructing the proper ultrasonic beam, this nonlinearity can be used to create, within the beam itself, an audible sound beam containing any sound desired. This is presently done in real-time using low cost circuitry, a specially designed amplifier, and transducers developed at MIT specifically for this project.

Hyper-directivity: The directivity, or narrowness, of an acoustic wave generated by a circular transducer is proportional to the ratio of the diameter of the transducer to the wavelength of the sound. So, a transducer much larger than the wavelength of the sound creates a very narrow beam.

Audible sound contains wavelengths reaching lengths of several feet, so a reasonably sized loudspeaker will always produce a very wide, non-directional source at lower frequencies. The Audio Spotlight, in contrast, outputs short, millimeter sized ultrasonic waves, which form a very narrow beam even in a small transducer, which in turn generates audible sound.

The nature of the nonlinear transformation also essentially eliminates side lobes in the resulting beam, and maintains relatively uniform directivity across the entire audible frequency range.

The figure above (from American Technologies Corp.) compares the directivity of the Audio Spotlight (yellow) to that of an ordinary loudspeaker (purple) at 400 Hz. Note that the directivity of the Audio Spotlight is only three degrees, compared to the essentially omni directional directivity of the loudspeaker.

In order to obtain such narrow directivity from a traditional loudspeaker system, one would need fifty meters across! A loudspeaker is like a light bulb, but the Audio Spotlight is like a laser.

History: The use of nonlinear interaction of high frequency sound to generate directive low frequency sound sources has been a well-researched subject in the field of underwater acoustics since the early 1960's. Often misattributed to so-called "Tartini Tones," the effect is more accurately described as a parametric array, a term introduced by Westervelt.

In the past several decades, many underwater sonar researchers have used the effect to generate directive low frequency sonar beams, detect underwater sound (parametric receiving array), and extend the bandwidth of underwater transducers.

The first published demonstration of an airborne parametric array was in 1975 by Bennett and Blackstock. Rather than using inaudible ultrasound, they instead used very intense, high frequency audible sound to produce simple difference tones. While their goal was not a practical audio reproduction device, they nonetheless effectively demonstrated that the parametric array would work in air in addition to underwater.

Life Assessment Detector System (LADS): The Life Assessment Detector System (LADS), a microwave Doppler movement measuring device, can detect human body surface motion, including heartbeat and respiration, at ranges up to 135 feet (41.15 meters).

The primary function of the LADS is to provide a reliable method by which medical and emergency personnel can locate personnel buried in building collapses or injured on the military battlefield. LADS can detect such signs of life as movement, heartbeat, or respiration.

Originally designed to detect heartbeat and respiration of military personnel wearing chemical-biological warfare protective over garments, the LADS have been restructured, greatly increasing its operational range and providing a means for eliminating "nuisance alarms" which could mimic human life signs, such as fans, wind drafts, or swaying trees.

This is accomplished through neural network technology, which "trains" the system to recognize human motion and heartbeat/respiration functions. If these functions are not detected, the reasonable assumption is that there are no survivors. Operating under such an assumption, the rescue team can now proceed without fear of further loss of life, i.e., rescue and medical personnel and equipment can be deployed more effectively and efficiently.

The LADS consist of a sensor module, a neural network module, and a control/monitor module. The sensor module is an x-band (10 GHz) microwave transceiver with a nominal output power of 15 milliwatts, operating in the continuous wave (CW) mode. The neural network module device can store many complex patterns such as visual waveforms and speech templates, and can easily compare input patterns to previously "trained" or stored patterns.

The control/monitor module provides the LADS' instrument controls, such as on-off switches, circuit breakers, and battery condition, as well as motion, heartbeat waveform, pulse strength, and pulse rate displays.

LADS provide life assessment capabilities for people who are:

- Trapped in building rubble;

- Battlefield casualties in a chemical/biological warfare environment;

- Victims of airline, train, or automobile crashes;

- Trapped in an avalanche or mudslide;

- Trapped on a mountain ledge;

- Trapped under a collapsed tent structure; or

- Hostages being held in a nonmetallic room.

For more information about the LADS, send E-mail to: info@vsecorp.com

Radar Flashlight: The National Institute of Justice (NIJ), through the Joint (Justice-Defense) Program Steering Group (JPSG), is sponsoring Georgia Tech Research Institute (GTRI) in developing inexpensive, handheld, low-power radar that will enable law officers to detect individuals through interior building walls. It works by sensing the motion of an individual's chest when they breathe.

GTRI is currently designing and refining the first prototype unit. A laboratory test area has been constructed consisting of a section of home siding and drywall, a wooden front door, and a section of brick and mortar.

It also demonstrated the ability to detect an individual through the laboratory's cinder block walls. GTRI is working to combine the two parts of this device into a single unit. NIJ plans on demonstrating the Radar Flashlight with law enforcement agencies through its National Law Enforcement and Corrections Technology Center (NLECTC) (Southeast Regional Center) before the end of 1999.

Dr. Pete Nacci Project Manager

Millivision Radar Millimeter-Wave Camera picks up both metallic and plastic concealed handguns. Between microwave and infrared lies the millimeter wave band. This little-heralded portion of the electromagnetic spectrum turns out to be perfect for "remote frisking." Millitech Corp. has designed a camera to accomplish just that.

The idea calls for measuring the time delay and intensity of millimeter wave energy that radiates naturally. At millimeter wavelengths, people are good emitters, while metals are very poor. Dielectric objects, such as plastics, ceramics and powdered drugs, are somewhere in between. Clothing and building materials, such as wallboard, are virtually transparent.

http://www.millivision.com/

Ground (or Home/Apt. Wall) Penetrating Radar: Patriot Scientific Corporation has developed radar technologies with a wide range of possible applications.

A pulse generator is used to drive the transmit antenna. The pulse is a positive spike going up to 100V then falling back to ground in one and a half nanoseconds corresponding to a pulse transmit frequency of 750 MHz.

The return signal is read by the receiving antenna. At this point some simple analog processing is done and the signal is digitized at a resolution of 6 GHz, and sent to a PC. The PC correlates the data into a conventional waveform, does some processing, and then transmits the data over an Ethernet cable to a Pentium workstation (not shown).

The Pentium workstation is used to apply different digital filters, combine waveforms, and display the results. This system can be used to demonstrate detection of small targets buried in sand, people behind walls, and other targets.

Patriot has used its antenna system to demonstrate detection of objects as small as a coke buried in sand, through a wall. Even small targets disturb the wave front of the pulse, producing reflections and modifying the field in measurable ways.

The key to Patriot's Radar system is its ability to transmit and receive pulses barely longer then single cycles at the transmit frequency. The first waveform shown here is a pulse generated by an earlier Patriot Design, based on "off the shelf" antenna technology. The waveform on the bottom was produced and received by Patriot's current Design.

The current Patriot antenna system produces a pulse at the desired frequency with little leading or trailing noise. The Patriot antenna system provides many advantages over pulse-based systems.

Patriot originally developed the impulse radar system to allow time domain processing in Patriot's GPR systems. Because the impulse is extremely short (3 nanoseconds), the time to return can be used to gauge the distance traveled by the pulse.

Furthermore, the transmitter and receiver antennas are very directional, eliminating much of the multipath components of the return signal. The short pulses combined with the directional transmit and receive to provide us with a number of important advantages:

- Very low average power during transmission
- Low interference from other transmitters
- Transmission invisible to conventional receivers
- High bandwidth digital data transmission possible
- Difficult detection by another impulse receiver

Interference with other sources and receivers is further reduced by using directional antennas. The antenna design shown is highly directional.

When penetrating the ground, we wish to eliminate as much of the multipath signal as possible. The directional antennas reduce the multipath signals detected to those that are relatively in line with the wave path, and eliminate much of the multipath signal that returns at odd angles.

Impulse radar uses low power inherently because the transmissions occur in pulses separated by periods of no transmission. The power of the pulses is offset by the dead time between the pulses. The average output of the current system is about 300 microwatts. The low average power of an impulse system effectively hides the transmissions from conventional receivers.

Interference can be further reduced in an impulse system by using random interval spacing. As long as the transmit and receive antennas are in sync, the period between pulses can be varied to prevent aliening with other continuous- or pulse-transmission systems that might be operating in the same locale.

Furthermore, if an impulse system is being used to transmit data, varying the intervals between pulses prevents other impulse systems from locking onto the signal. Patriot Scientifics' current GPR system does not use random interval spacing.

Patriot Scientific Corporation

Commercial Thought Reading Devices: The Cyberlink Mind Mouse is a revolutionary hands-free computer controller which allows you to move and click a mouse cursor, play video games, create music, and control external devices, all without using your hands.

A headband with three sensors detects electrical signals on the forehead resulting from subtle facial muscle, eye, and brain activity. This headband connects to an interface box that amplifies and digitizes the forehead signals and sends them to your computer.

The Cyberlink software decodes the forehead signals into ten Brain Fingers for continuous cursor control. It also decodes eye motion and facial gestures into mouse button clicks, keystrokes, and cursor resolution control. With a little practice, most or all of these commands can be mastered to operate virtually all computer functions.

By learning to change the energy levels of your Brain Fingers, you will be able to do just about anything on a computer, except turn it on! The Cyberlink Mind Mouse supports hands-free mouse, keyboard and joystick cursor control, switch closure, video game control, and music and art synthesis.

The Cyberlink Mind Mouse features a Windows 95 Mouse Driver for hands-free control of third party software like games, business software, Internet browsers, and a range of assistive technologies, such as the X-10 Home Controller and special needs word-processing and communication software, including WiVik2, Words Plus, and Clicker Plus.

The Cyberlink Mind Mouse is priced at $1495.00 (U.S.) plus shipping. Free upgrades are included for one year.

Hearing for the Deaf: It was during these [Frey] studies that a profoundly important discovery was made: Deaf subjects often had the ability to hear radio frequency sound. The clinical criterion was that, if a given person could hear audio above 5 kHz [higher range of a piano] by bone or air conduction, then radio frequency sound could be heard as well.

This and related work has resulted in the manufacture of radio frequency type hearing aids for the deaf, one of which is made by Listening, Inc., of Arlington, Mass., and is known as the Neurophone Model GPF-1. It operates at 100 kHz (about five times the normal maximum hearing frequency) and employs crystal control.

These observations tie in with the fact that some individuals can detect radio programs through fillings in their teeth. This phenomenon was technically verified by interposing shields between respective people who exhibited this effect and the modulated radio frequency sources.

When the lower half of the head was covered, including the maxillary dental area, the radio frequency sound was perceived. The sound ceased on covering the top half of the head. While the mechanism responsible for this phenomenon is only imperfectly understood, it can be assumed to be the result of DIRECT cortical stimulation.

In other words, even when the sound seems to be coming from the teeth, it is actually being directly received and interpreted in the brain, not the teeth.

Some Important Historical Developments

Brain-Wave Detection: Some 40-odd years ago, university professor F. Cazzamalli started publishing papers on the subject of brain-wave detection [using radio signals] and implied that he had detected radiations from the mind.

He placed his subjects in a shielded room (or Faraday cage), emanated VHF radio waves through their heads, and claimed to have recorded "beat frequencies" obtained with an untuned receiver consisting of a galena crystal or diode tube, a fixed capacitor, an antenna, and a sensitive light beam galvanometer. A "galvanometer" is a voltmeter; light beam types show up in physics labs and are one of the most sensitive types of voltmeter.

The trouble is that Cazzamalli never mentioned transmitter power in his somewhat unprofessional papers. His oscillograms meant to show variations of the "beat" when his subjects were emotionally aroused or engaged in creative tasks when they were in the Faraday cage. "Beat" as used by Cazzamalli refers to EEG-frequency, i.e., ELF, traces.

Later he told an astounded world that his subjects would hallucinate when under the influence of his "oscillatori telegrafica," its frequency being around 300 MHz at the time. Aviation radios are in this range.

Tom Jaski, a noted science writer and engineer duplicated some of Cazzamalli's work with a modern low-power oscillator that was swept from 300 MHz to 600 MHz. Cell phones start at over 900 MHz.

His subjects could not see the dial. They were told to sound off as soon as they felt something unusual. At a certain frequency range - varying between 380 MHz and 500 MHz - the subjects repeatedly indicated points with exact accuracy in as many as 14 out of 15 trials. At these "individual" frequencies, the same subjects announced having experienced pulsing sensations in the brain, ringing in the ears, and an odd desire to bite the experimenters.

The oscillator's output power was only a few milliwatts, while the oscillator itself was located several feet away from the subjects. Any experimenters out there want to try this? Milliwatts are quite safe for short-term experiments. Kids' walkie-talkies are 50 to 100 milliwatts, for example.

LIDA Machine: 1960 Soviet device that bombards brains with low-frequency radio waves. Now with Dr. Ross Adey, Chief of Research, Veterans Hospital, Loma Linda, CA.

Low frequency square wave modulation of a radio frequency field. It was developed by L. Rabichev and his colleagues in Soviet Armenia, for "the treatment of neuropsychic and somatic disorders, such as neuroses, psychoses, insomnia, hypertension, stammering, bronchial asthma, and asthenic and reactive disturbances." (U.S. Patent # 3,773,049)

The radio frequency field has a nominal carrier frequency of 40 MHz and a maximum output of approximately 40 watts. The E-field is applied to the patient on the side of neck through two disc electrodes approximately 10 cm in diameter. The electrodes are located at a distance of 2 to 4 cm from the skin.

The radio signal appears to be the primary cause of the sleep/trance effect. Optimal repetition frequencies are said to lie in the range from 40 to 80 pulses per minute.

Voice to Skull, 1974: The demonstration of sonic transduction of microwave energy by materials lacking in water LESSENS the likelihood that a thermo hydraulic principle is operating in human perception of the energy. Nonetheless, some form of thermo acoustic transduction probably underlies perception. If so, it is clear that simple heating is NOT a sufficient basis for the Frey effect; the requirement for pulsing of radiations appears to implicate a thermodynamic principle.

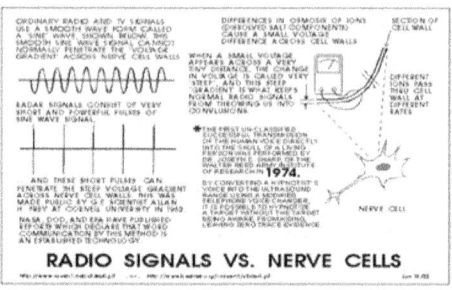

Frey and Messenger (1973) and Guy, Chou, Lin, and Christensen (1975) confirmed that a microwave pulse with a slow rise time is ineffective in producing an auditory response; only if the rise time is SHORT, resulting in effect in a square wave with respect to the leading edge of the envelope of radiated radio-frequency energy, does the auditory response occur.

Thus, the rate of change (the first derivative) of the waveform of the pulse is a CRITICAL factor in perception. Given a thermodynamic interpretation, it would follow that information can be encoded in the energy and "communicated" to the "listener".

Communication has in fact been demonstrated. A. Guy, a skilled telegrapher, arranged for his father, a retired railroad telegrapher, to operate a key, each closure and opening of which resulted in a pulse of microwave energy. By directing the radiations at his own head, complex messages via the Continental Morse Code were readily received by Guy.

Sharp and Grove found that appropriate modulation of microwave energy can result in "wireless" and "receiver less" communication of SPEECH. They recorded by voice on tape each of the single-syllable words for digits between 1 and 10. The electrical sine-wave analogs of each word were then processed so that each time a sine wave crossed ZERO REFERENCE IN THE NEGATIVE DIRECTION, a brief pulse of microwave energy was triggered.

This is, in effect, is a form of what is called pulse-RATE modulation. By radiating themselves with these "voice modulated" microwaves, Sharp and Grove were READILY able to hear, identify, and distinguish among the 9 words. Persons with artificial larynxes not unlike those emitted the sounds heard.

Communication of more complex words and of sentences was not attempted because the averaged densities of energy required to transmit longer messages would approach the current 10 milliwatts per square centimeter limit of safe exposure.

The capability of communicating directly with a human being by "receiver less radio" has obvious potentialities both within and without the clinic. But the hotly debated and unresolved question of how much microwave radiation to which a human being can safely be exposed will probably forestall applications within the near future.

EC-130E Commando Solo: Primary function is "Psychological operations broadcasts." Air Force Mission statement: Commando Solo conducts psychological operations and civil affairs broadcasts missions in the standard AM, FM, HF, TV and military communications bands.

It was called EC-121, known at the time as the Coronet Solo, and was used in Operation Urgent Fury. Volant Solo was during Operation Just Cause.

The Commando Solo aircraft and earlier generations of this design have participated in the following missions:

- Operation Urgent Fury (Grenada, October-November 1983, January-June 1985)

- Operation Just Cause (Panama, late December 1989)

- Operation Desert Shield (Kuwait, Iraq, from August 1990)

- Operation Desert Storm (Saudi Arabia, Turkey, Iraq, 1991)

- Operation Uphold Democracy (Haiti, 1994-1995)

- Operation Joint Guard (Part of a U.N. operation in Bosnia-Herzegovina, 1995)

- Operation Desert Thunder (part of a U.N. operation in Iraq)

- Operation Desert Fox (Iraq, 2 to 3 days in December 1998)

HAARP: HAARP (High Frequency Active Auroral Research Project), by virtue of its far-reaching impact on the environment to be a global concern and calls for its legal, ecological and ethical implications to be examined by an international independent body before any further research and testing, regrets the repeated refusal of the United States Administration to send anyone in person to give evidence to the public hearing or any subsequent meeting to be held by its competent committee into the environmental and public risks connected with the HAARP programmed currently being funded in Alaska.

One of HAARP's potential uses is a communications system. The military officially acknowledges two communications-related applications:

(1) to replace the existing Extremely Low Frequency (ELF) submarine communications system now operating in Michigan and Wisconsin;

(2) to provide a way to wipe out communications over an extremely large area, while keeping the military's own communications system working.

As we have seen, the mind-control subliminal messages are carried on radio-frequency broadcasts. The HAARP facility could be used to broadcast global mind-control messages, or such messages could simply be inserted into existing systems.

Dr. Igor Smirnov, of the Institute of Psycho-correction in Moscow, says in regard to this technology: "It is easily conceivable that some Russian 'Satan', or let's say Iranian [or any other 'Satan'], as long as he owns the appropriate means and finances, can inject himself [intrude] into every conceivable computer network, into every conceivable radio or television broadcast, with relative technological ease, even without disconnecting cables.

Media on HAARP – HAARP is set for 2015

HAARP TRANSMITTER NOW
RUNNING AT FULL POWER!!

Can be Easily Heard Around the World on Short-wave Radio. Has Space War Begun?

-BJ News by Marshall Smith

"As of this morning, Saturday Feb 17, 2001, HAARP began doing testing with greatly increased FULL power. The transmitter can now be heard all day long on 3.39 MHz. Very early this morning about 3 AM, HAARP could be heard at its "old" normal signal strength. About 4 AM the signal changed in both its pulse timing and inter-pulse spacing. At 4:30 AM the signal strength suddenly increased tremendously.

"Unlike previous mornings, there was no regular F-layer daytime fade out when the sun rose here in California about 6:45 AM. I continued to monitor during the daylight. I have never heard the HAARP signal during the daytime before. The sun now rises in Gakona Alaska about 10 AM PST. The received signal again increased from about S5 to S9 at 10:05 AM. With sunlight at both the transmitter and receiver there is no F-layer skip to bend the powerful signal around the planet. This means this is an extremely powerful direct ground wave signal. And I'm only receiving the leakage off the side lobes of the antenna array.

"The full HAARP design power is supposed to be about 350 Megawatts. But that is only the published spec, not necessarily what is done in practice (as in those CB'rs running illegal 1KWatt linear amplifiers). There is a planned Air Force "Star Wars" test with two vehicles, one from California and the other in the south Pacific, similar to last summer's failed test. The tentative published launch is set for late March or April. I will monitor HAARP to confirm it is running full power during the launch, as it was last summer.

"Last summer's "failure" is exactly what a HAARP device is supposed to do; destroy the electronic controls of a vehicle so the second stage cannot separate from the booster. A very cheap, simple way to knock down missiles launched from anywhere on the planet. It also can destroy military satellites in low orbit. Maybe that's why the Russians and Chinese have been complaining in the last several weeks about Bush's intention to "build" the Star Wars system. Maybe they've been losing some of their "secret hardware." But of course, they won't say that in public.

"It's now 11 AM PST, on Saturday the 17th, and the signal is blasting in with the powerful pre-pulse tone around S+20 and the main signal about S9. The signal varies 3 to 6 db over a series of several pulses. Since this is not due to F-layer skip fading, I must assume they are slewing the beam of the antenna in various directions, and thus changing the amount of the side lobes in this direction. This must be a test of a simulated space warfare game with multiple targets. Rapid slewing of the antenna in just a matter of a minute or two is not useful for submarine communication, nor for their stated purpose of doing "ionospheric research."

"To show the HAARP signal is abnormally large, at this time, the 80-meter band is silent and WWV at 5 MHz cannot be heard, as would be expected during the daytime. WWV at 10 MHz is barely heard but does not even register on the S meter. Tuning back to 3.39 MHz, the S meter jumps off the top of the scale. Even the extremely powerful Russian "woodpecker" transmitters during the cold war never did that, and they were aimed along the ground not out into space. I have no way to estimate how many Gigawatts that represents.

"It may be only coincidence but just several days ago, Russia announced it would be conducting a massive space war game, including the launch of numerous missiles, from both ground sites and submarines. Of course, this is only a coincidence. You Think. For more information about HAARP, how the transmitter works and to hear what the transmitter sounds like, go back to the Brother Jonathan Gazette front page article about the HAARP facility and how it is used in space warfare.

"I should point out in 1983 a number of Air Force ER-135 electronic warfare planes were shot down in the Sea of Japan. They were apparently making a covert entry into soviet airspace to test the latest Russian technology. What the Air Force did not know then was the Soviets had developed a stealth fighter so the 135's never saw the Russians coming and all 5 of the US e-warfare planes were shot down. To cover this "covert" event, the US shot down a 747, a plane similar to the 135's (or modified Boeing 707's whose parts are very similar to a 747) so if plane parts are found in the Sea of Japan they are claimed to be the 747. The 747 may, in fact, have played a part in the covert event. This is the supposed "Russian" shoot down of Korean Airlines Flight 007, on Sept 1, 1983. I remember the event well, since it is my son's b-day.

"The proof of a covert event with stealth Russian fighters shooting down 5 Air Force ER- 135's is documented in R.W. Johnson's book, SHOOTDOWN, published in 1986. The most convincing evidence is the strange fact that 27 US active duty electronic warfare officers somehow end up on the passenger list among the dead on the civilian KAL Flight 007 going to Korea. I only point this out to show how high-tech secret warfare between Russia and the US may result in deaths and the destruction of hardware, and yet is never reported to the public.

"The US did not announce and demonstrate deployment of its own stealth fighters until the Gulf War in '90-91, seven years later. In 1991, three events occurred, (1) the US demonstrates stealth fighter-bombers which can travel anywhere in the world without detection, (2) the announcement of the construction of HAARP which would neutralize all soviet missiles coming over the pole, and (3) the collapse of the Soviet Union. To see these as unrelated events is to miss the point of history.

"At the present time, both the Russian's and Chinese have demonstrated their ability and inclination to engage in warfare, especially space warfare. It would thus seem clear the "coincidence" of the massive Russian war games and the sudden increase in the output of HAARP in a warfare mode, would indicate that on this Presidents Day Weekend 2001, warfare is actually occurring, not just games. Just as in 1985, when planes were destroyed and US airmen died, the story was completely covered up, but it nonetheless had great implications on the relations between the governments of the world.

"There are, of course, no airmen on the Russian missiles, nor the Chinese and Russian satellites in orbit. This is the new hi-tech robotic remote-control warfare of outer space. But, the "war games" are real, nonetheless. The massive increase in the output of HAARP, under the control of the Air Force's Space Vehicles Command which operates HAARP and has the mission of engaging in space warfare, would indicate of a lot of expensive space hardware is now biting the dust. The Russians will claim their exercise was a "success." The Chinese who have just lost their "eyes in orbit" will say nothing. And the US will claim, as usual, "What, who me? HAARP hasn't been in operation since October '99." But you can listen for yourself on any short-wave receiver by tuning to 3.390 MHz. Good Listening."

Some Concluding Remarks

The discovery of "Synthetic Telepathy," as an EM way to communicate information directly onto the brain has changed everything. And, with related discoveries on membrane responses to specific frequency bands, and their amplification, has led to some very serious questions. Where are these technologies really leading man?

Certainly not toward a better way of living.... We see little application in their uses in medicine. In fact, most aspects have remained classified, and are still being used (historically) for military applications and uses. When you consider our ability to clone emotions, and the possible uses of HAARP, there is NO QUESTION on the possibility of its use for the complete mind control of humanity.

We can do it now. We have the theory. We have the technology. Why would it not be done? And then, by whom? Dr. Robert O. Becker, twice-nominated for the Noble prize for his health work in bio-electromagnetism, was more explicit in his concern over illicit government activity. He wrote of "obvious application in covert operations designed to drive a target crazy with voices."

What is frightening is that words, transmitted via low-density microwaves or radio frequencies, or by other covert methods, might be used to create influence. For instance, according to a 1984 U.S. House of Representatives report, a large number of stores throughout the country use high frequency transmitted words (above the range of human hearing) to discourage shoplifting. Stealing is reduced by as much as 80 percent in some cases.

With frequencies of electromagnetic radiation able to be converted into tight beams, it is worth considering the possibility that low-orbiting satellites could transmit such signals directly to a chosen person. Such a capability would have obvious strategic value.

What if some group or organization wanted to conduct terrorist activities that would strike right into the heart of their intended victims? What better way than to create killers from the most ordinary of citizens? Worse yet would be to cause children to unexpectedly lash out and murder those around them.

It is apparent that the technology exists to elicit such behavior using waves of electromagnetic energy sent directly into the brain.

"ELF" frequencies, as we know now, are not the most invasive weapons-capable frequencies. Doses of ELF can act as either sedatives or stimulants, entraining the target's EEG as with the Russian LIDA machine. Weapons that can transmit hypnotic commands silently and untraceably, over distance and through walls, can cause a far wider spectrum of effects.

In other words, don't assume that "ELF" is the only form of electromagnetic signal which can do damage. And, if you might wonder what that "buzzing" sound you sense behind you ear might be? It is HAARP. You can actually hear it when they have it "on."

The obvious next important discussion should be around the changing Shuman Resonance, and its affects and importance to biological functions and activities. There is no question that the Shuman Resonance is changing toward the higher frequencies associated with our "conscious states" of mind... Why is that?

View Book and Media References (currently under editing)

IMPORTANT TECHNICAL REFERENCES

- http://www.bibliotecapleyades.net/sociopolitica/esp_sociopol_mindcon30b.htm

- Photo and description of the Korean War LIDA machine, a radio frequency brain entrainment device.

- "Human Auditory System Response to Modulated Electromagnetic Energy," Allan H. Frey, General Electric,

Advanced Electronics Center, Cornell University, Ithaca, New York.

- NASA technical report abstract stating that speech-to-skull is feasible.

- DOD/EPA small business initiative (SBIR) project to study the unclassified use of voice-to- skull technology for military uses.

- Excerpts, Proceedings of Joint Symposium on Interactions of Electromagnetic Waves with Biological Systems, 22nd General Assembly of the International Union of Radio Science, August 25 - September 2, 1987, Tel Aviv, Israel.

- Excerpt, Dr. Don R. Justesen, neuropsychological researcher, describes Dr. Joseph C. Sharp's successful transmission of "words" via a pulse-rate- modulated microwave transmitter of the Frey type.

- FOIA article circulated among U.S. agencies describing the Russian TV program "Man and Law", which gives a glimpse into the Russian mind control efforts. (Dr. Igor Smirnov, a major player, was used as a consultant to the FBI at the Waco Branch Davidian standoff.) SBIR (small business initiative contract) which clearly shows intent to use ultrasound as an anti-personnel weapon, including one-man portability and with power to kill.

- A page originally from the MIT Media Lab's acoustic engineer, Joseph Pompeii. Describes a similar technique under commercial and military development (American Technologies Corp., San Diego) under the trade name "Hypersonic Sound". Shows that sound can be focused to the extent of targeting just one person in a crowd, acoustically, using ultrasound.

APPENDIX A

ELECTROMAGNETIC WEAPONS TIMELINE:

— Electromagnetic (EM) weapons are of recent invention. They utilize the various frequencies of the electromagnetic spectrum to disable or kill the target. Psychotronic weapons are those EM weapons that interact with the nervous system of the target. These weapons usually operate in the very low (100 to 1,000 Hz) or extremely low (greater than zero but less than 100 Hz) frequency ranges.

— 1934 "A method for Remote Control of Electrical Stimulation of the Nervous System", a monograph by Drs. E. L. Chaffee and R. U. Light.

— 1934 Experiments in Distant Influence, book by Soviet Professor Leonid L. Vasiliev. Vasiliev also wrote an article, "Critical Evaluation of the Hypogenic Method" concerning the work of Dr. I. F. Tomashevsky on experiments in remote control of the brain.

— 1945 After World War II, the Allies discovered the Japanese had been developing a "death ray" utilizing very short radio waves focused into a high-power beam. Tests were done on animals.

— 1950 The French conducted research on infrasonic weapons. (From "The Road from Armageddon", by Peter Lewis, Resonance #13, pp 9-14).

— 1953 John C. Lilly, when asked by the director of the National Institute of Mental Health (NIMH) to brief the Central Intelligence Agency (CIA), Federal Bureau of Investigation (FBI), National Security Agency (NSA), and the various military intelligence services on his work using electrodes to stimulate directly the pleasure and pain centers in the brain, refused.

- 1958, 1962 The U.S. conducts high-altitude Electromagnetic Pulse (EMP) bomb tests over the Pacific. (From "The Road from Armageddon" by Peter Lewis.)

- 1960 Headlines read "Khrushchev Says Soviets Will Cut Forces a Third; Sees 'Fantastic Weapon' ". (From article of same title, by Max Frankel, New York Times, Jan. 15, 1960, p1 as cited in "Tesla's Electromagnetics and Its Soviet Weaponization", paper by T. E. Bearden.)

- 1965 A "Death Ray" weapon was developed by McFarlane Corporation, described as a modulated electron gun X-ray nuclear booster, could be adapted to communications, remote control and guidance systems, EM radiation telemetry and death ray.

- Reported hearings before the House Subcommittee on Department of Defense Appropriations, chaired by Rep. George Mahon (Dem. - Texas). From "Hearing Voices" by Alex Constantine, Hustler, January 1994, pp. 102-104, 113, 120, 134. Research by Harlan Girard.

- 1965 Project Pandora was undertaken in which chimpanzees were exposed to microwave radiation. The man who was in charge of this project said, 'the potential for exerting a degree of control on human behavior by low level microwave radiation seems to exist' and he urged that the effects of microwaves be studied for 'possible weapons applications' - (From "Electromagnetic Pollution: A Little Known Health Hazard. A new means of control?" by Kim Besley, Great Britain, p14. Research from Woody Blue).

- 1968 Dr. Gordon J. F. MacDonald, science advisor to President Lyndon Johnson, wrote, "Perturbation of the environment can produce changes in behavioral patterns." He was referring to low frequency EM waves in the ionosphere affecting human brain wave patterns. (From his book, Unless Peace Comes, a Scientific Forecast of New Weapons, cited in "New World Order ELF Psychotronic Tyranny", a paper by C. B. Baker.)

— 1970 Zbigniew Brzezinski, President Jimmy Carter's National Security Director, said in his book, Between Two Ages, weather control was a new weapon that would be the key element of strategy. "Technology will make available to leaders of major nations a variety of techniques for conducting secret warfare..." He also wrote "Accurately timed, artificially excited electronic strokes could lead to a pattern of oscillations that produce relatively high power levels over certain regions of the Earth ... one could develop a system that would seriously impair the brain performance of a very large population in selected regions over an extended period."(Cited in Baker's "ELF Psychotronic Tyranny" paper.)

— 1972 The Taser, first electrical shock device developed for use by law enforcement, delivers barbed, dart shaped electrodes to a subject's body, and 50,000 volt pulses at two millionths of an amp over 12-14 seconds time. (From "Report on the Attorney General's Conference on Less Than Lethal Weapons", by Sherry Sweetman, 1987, p4, which cites "Non-Lethal Weapons for Law Enforcement: Research Needs and Priorities. A Report to the National Science Foundation by the Security Planning Corporation, 1972. Research by Harlan Girard.)

— 1972 A U.S. Department of Defense document said that the Army has tested a microwave weapon. It was an extremely powerful 'electronic flamethrower'." (From "Electromagnetic Pollution.")

— 1972 A study published by the U.S. Army Mobility Equipment Research and Development Center, titled 'Analysis of Microwaves for Barrier Warfare' examines the plausibility of using radio frequency energy in barrier counter-barrier warfare ... The report concludes that:

a) it is possible to field a truck-portable microwave barrier system that will completely immobilize personnel in the open with present day technology;

b) there is a strong potential for a microwave system that would be capable of delaying or immobilizing personnel in vehicles;

c) c) with present technology, no method could be identified for a microwave system to destroy the type of armored material common to tanks."

— (From "Electromagnetic Pollution" by Kim Besly, p15, quoting The Zapping of America by Paul Brodeur.) The report further documents the ability to create third-degree burns on human skin using 3 GHz at 20-watts/square centimeter in two seconds.

— 1972 Dr. Gordon J. F. MacDonald testified before the House Subcommittee on Oceans and International Environment, concerning low frequency research: "The basic notion there was to create between the electrically charged ionosphere in the higher part of the atmosphere and conducting layers of the surface of the Earth this neutral cavity, to create waves, electrical waves that would be tuned to the brainwaves ... about ten cycles per second ... you can produce changes in behavioral patterns or in responses." (From Baker's "ELF Psychotronic Tyranny" paper.)

— 1973 Sharp and Grove transmit audible words via microwaves.

— 1975 - 1977 "Unpublished analyses of microwave bio-effects literature were disseminated to the U.S. Congress and to other officials arguing the case for remote control of human behavior by radar." (From the Journal of Microwave Power, 12(4), 1977, p320. Research by Harlan Girard.)

— 1978 Hungarians presented a state-of-the-art paper on infrasonic weapons to the United Nations, "Working Paper on Infrasound Weapons", United Nations CD/575, 14 August 1978. (From "The Road from Armageddon" by Peter Lewis.)

— 1981 - 1982 "Between 1981 and September 1982, the Navy commissioned me to investigate the potential of developing electromagnetic devices that could be used as non-lethal weapons by the Marine Corp for the purpose of 'riot control', hostage removal, clandestine operations, and so on." Eldon Byrd, Naval Surface Weapons Center, Silver Spring MD. (From "Electromagnetic Pollution" by Kim Besly, p12.)

— 1982 Electromagnetic weapons for law enforcement use in Great Britain: Charles Bovill of the now defunct British firm, Allen International, developed a 10-30 Hz strobe light that can produce seizures, giddiness, nausea, and fainting. Addition of sound pulses in the 4.0 - 7.5 Hz range increases effectiveness, as utilized in the Valkyrie, a "frequency" weapon advertised in British Defense Equipment Catalogue until 1983.

— The squawk box or sound curdles uses two loudspeakers of 350-watt output to emit two slightly different frequencies that combine in the ear to produce a shrill shrieking noise. The U.S. National Science Foundation report says there is "severe risk of permanent impairment of hearing." (From "Electro-pollution" by Kim Besley, citing the Manchester City Council Police Monitoring Unit document.)

— 1982 Air Force review of biotechnology: "Currently available data allow the projection that specially generated radio frequency radiation (RFR) fields may pose powerful and revolutionary antipersonnel military threats. Electroshock therapy indicates the ability of induced electric current to completely interrupt mental functioning for short periods of time, to obtain cognition for longer periods and to restructure emotional response over prolonged intervals.

— "... impressed electromagnetic fields can be disruptive to purposeful behavior and may be capable of directing and/or interrogating such behavior. Further, the passage of approximately 100 milliamperes through the myocardium can lead to cardiac standstill and death, again pointing to a speed-of-light weapons effect.

— "A rapidly scanning RFR system could provide an effective stun or kill capability over a large area."

Remote Brain Targeting

- (From Final Report on Biotechnology Research Requirements for Aeronautical Systems Through the Year 2000. AFOSR-TR-82-0643, Vol 1, and Vol 2, 30 July 1982.)

- 1986 "The Electromagnetic Spectrum in Low-Intensity Conflict" by Captain Paul Tyler, MC, USN quotes the above passage and further elaborates on the theme. (Published in Low Intensity Conflict and Modern Technology Lt. Col. David J. Dean, USAF, ed., Air University Press, Maxwell AFB, AL. Research by Harlan Girard.)

- 1983 Nikolai Khokhlov, a Soviet KGB agent who defected to the West in 1976, interviews recently arrived scientists and reports: "The Soviet mind- control program is run by the KGB with unlimited funds." (From The Spectator, February 5, 1983, reported in "New World Order Psychotronic Tyranny" by C. B. Baker.)

- 1984 "USSR: New Beam Energy Possible?" possibly associated with early Soviet weather engineering efforts over the U.S. (From "Tesla's Electromagnetics and Its Soviet Weaponization" by T. E. Bearden.)

- 1985 Women in the peace camps at Greenham Common began showing various medical symptoms believed to be caused by EM surveillance weapons beamed at them. (See "Zapping: The New Weapon of the Patriarchy", Resonance #13, pp. 22-24. Research by Woody Blue.)

- 1986 Attorney General's Conference on Less Than Lethal Weapons reviews current weapons available. They include the Taser, the Nova XR-5000 Stun Gun (can interrupt a pacemaker); the Talon, a glove with an electrical pulse generator; and the Source, a flashlight with electrodes at the base.

- Photonic driving strobe lights tested by one conference delegate on 100 subjects, produced discomfort. Closed eyelids to not block the effect. Evidence that ELF produces nausea and disorientation. Suggestion to develop fast acting electro sleep inducing EM weapon.

— Discusses problem of testing weapons on animals and human "volunteers". (From "Report on the Attorney General's Conference on Less Than Lethal Weapons", by Sherry Sweetman, March 1987, prepared for the National Institute of Justice. Research by Harlan Girard.)

— 1988 The Pentagon is ordered by courts to cease EMP tests at several locations due to a lawsuit filed by an environmental group. (From The Washington Post, May 15, 1988, see "U.S. and Soviets Develop Death Ray", Resonance 11, p10. Research by Remy Chevalier.)

— 1992 December. "The U.S. Army's Armament Research, Development and Engineering Center is conducting a one-year study of ACOUSTIC BEAM TECHNOLOGY ... the command awarded the one-year study to Scientific Applications and Research Associates of Huntington Beach CA. Related research is conducted at the Moscow based Andreev Institute." (From "U.S. Explores Russian Mind Control Technology", by Barbara Opal, Defense News, January 11-17, 1993. Research by Harlan Girard and others.)

— 1993 The Russian government is offering to share with the United States in a bilateral Center for Psycho-technologies the Soviet mind-control technology developed during the 1970s. The work was funded by the Department of Psycho-Correction at the Moscow Medical Academy.

— "Acoustic psycho-correction involves the transmission of specific commands via static or white noise bands into the human subconscious..." The Russian experts, among them former KGB General George Kotov, present in a paper a list of software and hardware available for $80,000. (From Opal article, "U.S. Explores Russian Mind Control Technology".)

— 1993 February 28, beginning of 51 day siege on the Branch Davidians at Waco Texas, which ended in the death of more than 80 people.

— Until this incident, the electromagnetic weapons had kept a very low profile. But in the documentary video, "Waco: The Big Lie Continues", footage from the British Broadcasting (BBC) shows at least three EM weapons used by U.S. government agents. First, the noise generators used against the Davidians. Second, a powerful strobe light, shown during a nighttime sequence.

— And the third was the Russian psychoacoustics weapon, considered, but agents deny use of this weapon against the Waco people. FBI agents met with Dr. Igor Smirnov in Arlington VA to discuss the possibility of using the weapon against the Davidians. (From "A Subliminal Dr. Strangelove", by Dorinda Elliot and John Barry, Newsweek, August 22, 1994)

APPENDIX B

Most Common Known ELF Effects

Here is a list of most of the common effects. It is intended to show how the various induced stress effects are broken down. Indent levels are used to show categories and sub-categories:

— Invasive At-a-Distance Body Effects (including mind):

— Sleep deprivation and fatigue:

— Silent but instantaneous application of "electronic caffeine" signal, forces awake and keeps awake

— Loud noise from neighbors, usually synchronized to attempts to fall asleep

— Precision-to-the-second "allowed sleep" and "forced awakening"; far too precise and repeated to be natural

— Daytime "fatigue attacks", can force the victim to sleep and/or weaken the muscles to the point of collapse

— Audible Voice to Skull (V2S):

— Delivered by apparent at a distance radio signal

— Made to appear as emanating from thin air

— Voices or sound effects only the victim can hear

— Inaudible Voice to Skull (Silent Sound):

— Delivered by apparent at a distance radio signal;

— Manifested by sudden urges to do something/go somewhere you would not otherwise want to

- Silent (ultrasonic) hypnosis presumed

- Programming hypnotic "triggers" - i.e., specific phrases or other cues which cause specific involuntary actions

- Violent muscle triggering (flailing of limbs):

- Leg or arm jerks to violently force awake and keep awake

- Whole body jerks, as if body had been hit by large jolt of electricity

- Violent shaking of body; seemingly as if on a vibrating surface but where surface is in reality not vibrating

- Precision manipulation of body parts (slow, specific purpose):

- Manipulation of hands, forced to synchronize with closed-eyes but FULLY AWAKE vision of previous day; very powerful and coercive, not a dream

- Slow bending almost 90 degrees BACKWARDS of one toe at a time or one finger at a time

- Direct at-a-distance control of breathing and vocal cords; including involuntary speech iv.

- Spot blanking of memory, long and short term

Reading said-silently-to-self thoughts:

- Engineered skits where your thoughts are spoken to you by strangers on street or Real time reading sub vocalized words, as while the victim reads a book and BROADCASTING those words to nearby people who form an amazed audience around the victim

- Direct application of pain to body parts:

- Hot-needles-deep-in-flesh sensation

- Electric shocks (no wires whatsoever applied)
- Powerful and unquenchable itching, often applied precisely when victim attempts to do something to expose this "work"
- "Artificial fever", sudden, no illness present vs. sudden racing heartbeat, relaxed situation
- Surveillance and tracking:
- Thru-wall radar and rapping under your feet as you move about your apartment, on ceiling of apartment below
- Thru-wall radar used to monitor starting and stopping of your urination - water below turned on and off in sync with your urine stream
- Loud, raucous artificial bird calls everywhere the victim goes, even into the wilderness
- Invasive Physical Effects at a Distance (non-body):
- Stoppage of power to appliances (temporary, breaker ON)
- Manipulation of appliance settings
- Temporary failures that "fix themselves"
- Flinging of objects, including non-metallic
- Precision manipulation of switches and controls
- Forced, obviously premature failure of appliance or parts
- External Stress-Generating "Skits":
- Participation of strangers, neighbors, and in some cases close friends and family members in harassment:
- Rudeness for no cause

- Tradesmen always have "problems", block your car, etc.
- Purchases delayed, spoiled, or lost at a high rate
- Unusually loud music, noise, far beyond normal
- Break-ins/sabotage at home:
- Shredding of clothing
- Destruction of furniture
- Petty theft
- Engineered failures of utilities
- Sabotage at work:
- Repetitive damage to furniture
- Deletion/corruption of computer files
- Planting viruses which could not have come from your computer usage pattern
- Delivered goods delayed, spoiled, or lost at a high rate
- Spreading of rumors, sabotage to your working reputation
- Direct sabotage and theft of completed work; tradesmen often involved and showing obvious pleasure
- In summary, the effects include:
- To the Brain:
- Forced memory blanking and induced erroneous actions
- Induced changes to hearing. Both apparent direction and volume, and sometimes even content

- Reading and broadcasting thoughts. "How can that lady talk with her mouth closed?"

- Vivid controlled dreams.

- Forced waking visions: some synched with forced body motion

- Microwave hearing

- Sleep prevention each night, at exactly the same time (for weeks)

- To the Face:

- Sudden violent itching inside eyelids

- Forced manipulation of airways, including externally controlled forced speech

- "Transparent eyelids

- Artificial tinnitus (ringing in ears)

- Forced movement of jaw and clacking of teeth

- To the Body:

- Wilding racing heart without cause

- Remotely induced violent no-rash itching

- Forced nudging of arm during delicate or messy work causing injury or spills

- Forced "muscle quaking" on large muscles on the back or other unexplainable vibrations

- Cases of repeated fresh watch batteries dying

- Forced precision manipulation of hands. Sometimes synched to the forced waking visions.

- Special attention to genital area: itching, forced orgasm, intense pain, "hot needles"
- ntense general pain in the legs, like stings.
- [Remind you, that these are the unclassified effects only.]

CHAPTER FOUR

"There was truth and there was untruth, and if you clung to the truth even against the whole world, you were not mad."

- George Orwell (1984)

John J. McMurtrey, a biologist, is an individual well versed in the subject of electromagnetic technology and its connection and capability as a foundation for misdiagnosis of Targeted Individuals.

Through extensive scientific research and numerous studies based on Mind Control, Open Literature Evidence, McMurtrey concludes, beyond a reasonable doubt, the subliminal capability of delivery systems capability to deploying telepathic microwaves as extremely low frequencies (ELF) or what we have learned is the "Frey Effect." Many of McMurtrey's conclusions reveal in detail similarities and traits which can materialize as Schizophrenic symptoms after a target is, or has been, entered into a Remote Brain Targeting program. In reality, the target is being electronically harassed by not only microwave weaponry, but could be harassed also by other forms of technology for example, patented technology capable also of reading a person's mind. Below are excerpts:

MICROWAVE BIOEFFECT CONGRUENCE WITH SCHIZOPHRENIA

**Christians Against Mental Slavery
For complete details see:**

http://www.slavery.org.uk/science.htm

John J. McMurtrey M. S., Copyright 2002, 10 Apr. 2005
Co-authorship is negotiable towards professional publication in an NLM indexed journal,

Email- Johnmcmurt@aol.com

Donations toward future research are gratefully appreciated at:

http://www.slavery.org.uk/FutureResearch.htm

ABSTRACT

The substantiation for microwave voice transmission development, which can be isolated to an individual, prompts review of the correlation between microwave bio-effects and schizophrenia. These correlations are extensive. Studies of both conditions report short-term and spatial memory deficit, time estimation changes, deficits in sequencing, coordination deficit, numerous electro-physiologic changes, startle decrease, neurotransmitter changes, hormone alterations, immune alterations, mitochondria deficits, lipid phosphorylation decrease, lipid peroxidation, deleterious histologic change in disease reduced brain areas, activation of hallucination involved brain areas, and ocular disease. Schizophrenia findings correlate with microwave bio-effects so extensively as to indicate a congruence, and appear to implicate a microwave involvement with enough patients to be remarkable in study results. The development of methods to exclude microwave means in psychosis is imperative, and research is proposed.

INTRODUCTION

Remote microwave induced sound and internal voice technology has long been discovered, developed, detailed in patents, with weapons applications described. that such technology can be applied remotely and coupled to target tracking technology has implications for patients who, by virtue of voice transmission complaint and other symptoms, are diagnosed with various mental disorders. Auditory hallucination is most prevalent in schizophrenia, which features in 60% of cases. A frequent patient understanding of the origin of voices is by remote transmission, though the very concept is considered delusional, and often the diagnosis is psychosis of varying severity depending on functional ability, without any investigation of described internal voice capabilities.

The substantiation of microwave voice transmission development suggests examination of any microwave bio-effect correlation with schizophrenia findings. The hypothesis tested was that perhaps some discrepant schizophrenia study results could differentiate patients subjected to technological assault. Unfortunately, little differentiation was evident, because the correlations appear too extensive, as presented in overview. Table I. Unless otherwise noted, the microwave exposure effects examined are at low intensity, and are expressed in text parenthesis in terms of existing exposure standards. Since most of the observed correlations are close to microwave exposure standards, the possibility of an environmental microwave association with schizophrenia is considered.

Project Bizarre Weapons Implications: Are Psychiatric Diagnosis, and Microwave Exposure Standards Presumptive?

John J. McMurtrey, Copyright 2008,

10 Nov 2008

ABSTRACT

Reviewed are Freedom of Information Act disclosures for Project Bizarre, a monkey microwave exposure investigation prompted by Soviet irradiation of the American Embassy in Moscow, and further microwave exposure literature corroborating timed response or task sequencing disruption. Literature authenticating microwave hearing voice transmission development is also covered as well as schizophrenia time estimation and sequencing performance deficit reports similar to microwave exposure findings. Current medical practice is presumptive in being without knowledge or investigation of technological development relevant to patient complaints and those correspondences compared. Present radio frequency exposure standards are indicated inadequate regarding the parameters considered.

INTRODUCTION

The microwave irradiation of the American Embassy in Moscow received little publicity until the winter of 1976 instillation of protective screening, but irradiation was known since 1953. The irradiation was directional from nearby buildings with pulsation detected. Complaint to the Soviets had little avail, but the signals disappeared in January

1979 "reportedly as a result of a fire in one or more of the buildings," though there was recurrence in 1988. Psychiatric cases occurred during the exposure period, but no epidemiologic relationship was revealed with fully a quarter of the medical records unavailable, and comparison with other Soviet Bloc posts. Although significant results matched the Soviet recognized neurotic syndrome, these were dismissed as subjective symptoms. Professional publications further detail some of these flaws, along with charges of government cover-up, particularly respecting cancer cases. The Central Intelligence Agency had Dr. Milton Zaret review Soviet medical microwave literature to determine the purpose of the irradiation. He concluded the Russians "believed the beam would modify the behavior of the personnel. In 1976 the post was declared unhealthful and pay raised 20%.

The Soviet irradiation of the American Embassy prompted a 1965 White House directive to investigate radio frequency biological effects particularly in the microwave region, that resulted in a major classified project code named Pandora. Project Pandora became a number of subprojects, one of which was a rhesus monkey investigation dubbed Project Bizarre that was conducted by Dr. J. C. Sharp and H. M. Grove who later are noted to have developed a method for remotely transmitting intelligible words by modulations of the microwave hearing effect. Here reviewed are Project Bizarre microwave exposure results as know from previously classified Freedom of Information Act (FOIA) releases along with corroborating journal studies observing the same or similar deficits. Since radio frequency voice transmission implemented to simulate hallucination would involve analogous microwave exposures, those extant references to such development are as well covered with corresponding deficits observed in schizophrenia, which is the most well studied diagnosis containing numerous remote voice transmission complainants.

Project Bizarre

Project Bizarre involved designing a facility capable of uniformly irradiating primates that required an operational manual. Preliminary results indicating an effect on the monkeys' ability to perform operant tasks were reported to the Advanced Research Projects Agency Director, **Error! Bookmark not defined.** which were confirmed and yielded Director Memo stating: "The potential of exerting a degree of control on human behavior by low level selectively modulated microwave radiation should be investigated for potential weapons applications." **Error! Bookmark not defined.** By 1969, an Institute for Defense Analysis panel unanimously found degradation in monkey performance at 1 mW/cm2 up to 4.6 mW/cm2 on more than 10 days of 10 hours per day exposure. The Bizarre Project exposures simulated a signal of particular concern occurring at the American Embassy in Moscow that was in the "S" band centered about 3 GHz, but "L" band frequency also had occurrence, and onsite radiation levels inside the embassy in 1965 was "measured at values in excess of 1 mW/cm2" The 'Moscow Signal' simulation was quite complex, but significant to later literature is that the frequencies investigated were centered about 3 GHz, and there were two superimposed signals each effectively pulsed at 440 times per second. Then or presently, such results below US exposure standards question substantial military and commercial investment. The administration changed in 1969, and as of 1967 radiation levels measured at the Embassy had considerably decreased in power to "always below 50 microwatts/cm." By 1970, a Bizarre Project advisory committee member analyzed monkey performance data for microwave exposure duration without unexposed comparison that could not show behavioral differences and a report resulted examining the issues. In letter to the Advanced Sensors Director, conclusion emphasis was that "no evidence of any permanent, deleterious effects is to be expected," as no monkey performance degradation after exposure recovery was detectable. In July 1970, the project was relegated to the Walter Reed Army Institute of Research where the facilities were installed.

Microwave exposure produced monkey operant behavior changes of decreased ability to delay response for 50 seconds during an inter-response time (IRT) portion of a food reward schedule. The Bizarre Project primary investigator suggested the results be independently replicated, and would survive what most scientists consider invalid analysis. Scientists often of some military laboratory affiliation have produced very similar performance degradations to those found for Project Bizarre over 5 rat investigations at 3 different laboratories without any complex microwave signal modulation. A cell phone study indicates alteration of time perception in humans. Project Bizarre also found decreased ability in sequential tasks later covered before confirming journal literature review.

Technological Simulation of Hallucination
John J. McMurtrey, M. S. and Edward A. Moore, M. D.

Copyright 2006, 30 May 2011

ABSTRACT

Objective: Evidence for technologies capable of remote sound or voice transmission isolated to individuals is surveyed along with target tracking capacity that can maintain apparent psychosis. Method: Examination of government reports, engineering databases, the patent database post 1976, PubMed, and the Internet for available pertinent authentic sources. Results: Ultrasound and radio frequency methods are described to remotely isolate voice to individuals. Accounts of ultrasound and radio frequency energy forms used on people also exist. Conclusion: Evidence indicates development of technologies capable of remotely isolating sound and voice to an individual. Covert misuse of such technologies would result in simulated hallucination, which has no diagnostic recognition.

"Such a device has obvious applications in covert operations designed to drive a target crazy with 'voices' or deliver undetectable instructions to a programmed assassin."

--Robert O. Becker regarding microwave hearing voice transmission, who was twice Nobel Prize nominated for biological electromagnetic fields research.

INTRODUCTION [1]

Medical professionals regard the perception of voice or sound, which cannot be heard by others nearby as hallucination excepting only tinnitus, and deem such phenomena as psychotic manifestations

on persistent, disturbing complaint. Though 'hearing voices' can involve numerous diagnoses, this symptom is often considered characteristic in schizophrenia with 47-98% prevalence, but the symptom has reported prevalence in dissociative identity disorder (DID) of 30-64%, and in bipolar disorder of 7-48%. Patients frequently believe that such voices are externally transmitted to them. Apparently unrecognized by the psychiatric community, two technologies have the described capacity to remotely transmit voice or sound in an individually isolated manner. Considering that misuse of such technologies could simulate hallucination and confuse diagnosis, the available body of evidence for such capacity is reviewed.

Abbreviations: ABR = Auditory Brainstem Response. DID = Dissociative Identity Disorder. FOIA = Freedom of Information Act. Hz = cycles per second, and is an eponym honorific abbreviation for Heinrich Rudolph Hertz. LRAD = Long Range Acoustic Device. MHz = Mega-Hertz denoting one million cycles per second.

Herein is substantiated:

Development of remote ultrasound and radio frequency technologies for transmitting sound or voice, which can be isolated to an individual.

1. Human tracking technologies.

2. Reports and published anecdotes of ultrasound and radio frequency energy use against people.

METHODS

Literature examination by relevant terminology was performed on PubMed, National Technical Information Service, Google search, US and European patent office databases, as well as Compendex, the Wilson Web, and Inspec. Ultrasound and microwave bioeffects references were also cross examined for relevance per article. Inclusion/exclusion criteria are pertinence and authenticity.

Time series bibliometry of citations from the most recent radio frequency hearing review was plotted in histogram with differentiation of known military support (Figure 1). The significance of difference in total publication rate was assessed by t-test of means for baseline publication prior to 1972 from the 1961 Frey substantiation compared with the 1973-1980 apparent publication upslope (the 1956 citation is only an advertisement mentioning microwave hearing with omission increasing conservative comparison). Regression was plotted for the baseline with no upward trend, and per the concerns of single case time series analysis, the 1973-1980 upslope data were adjusted by per datum subtraction of the mean from Frey 1961 through 1972 publication baseline for the presented p value.

ULTRASOUND VOICE TRANSMISSION

The loud, steady production of two different tones results in a third tone equal to the frequency difference between the original tones. The sounds so created are known as the tones of Tartini who was an 18th century violinist, and result from air non-linearity that causes sound to scatter itself. The effect also occurs in water for sonar generators called parametric arrays, with the short ultrasound wavelength permitting high directional projection. Acoustic tones produced in air by ultrasound beams were first reported in 1962, followed by several abstract reports, and then had more complete publication. Voice modulated on an ultrasound beam is caused to peel off by another ultrasound beam in loudspeakers for directionally projecting sound, with mathematical prediction compared to experimental results. Basic methods for such speakers are described in the Audio Engineering Handbook. Recently reported are improvements in emitters, and directivity. Though utilized as a term in many reports, 'loudspeaker' has somewhat misleading connotations for these speakers, since virtual point sound sources are generated within the ultrasound beams without scattering outside the beam intersection. A recipient perceives this sound projection technique as originating within the head without directional orientation as described from demonstrations for an audio engineering society, an engineering news article, and Popular Science as well as non-lethal weapon applications patents.

An ultrasound voice transmission patent discusses non-lethal weapon use against crowds or as directed at an individual. Communication that is understood as an inner voice can have powerful emotional reactions in people, "since most cultures attribute inner voices either as a sign of madness, or as messages from spirits or demons. Another ultrasound voice transmission patent describes sound production particularly within cavities such as the ear canal. An individual readily understands communication by the device across a noisy crowded room without discernment by others nearby. Sound can also be made to appear as originating from mid-air or from surfaces by reflection.

American Technology Corporation licensed this latter patent, and commercially sells their Hyper Sonic Sound system, which has a technical treatment available and a professional meeting presentation. This company also has an acoustic non-lethal weapon called the Long Range Acoustic Device (LRADTM) that is deployed to the Navy, Coast Guard, Army, Marine Corps, military prison camps, and the US Border Patrol as well as ground troops in Iraq and Afghanistan. An 80 % efficacy in deterring wayward Persian Gulf vessels by the LRAD has science news report. The device is also deployed to police departments, **Error! Bookmark not defined.** cruise ships, **Error! Bookmark not defined.** and at petroleum instillations, while a version of the device is available for automatic operation in conjunction with remote sensor security systems. A similar ultrasound method capable of limiting sound to one person, Audio Spotlight® has peer reviewed publication, **Error! Bookmark not defined.** and is marketed. Audio Spotlight press releases indicate exhibition at Boston's Museum of Science, the General Motors display at Disney's Epcot Center, the Smithsonian National Air & Space Museum, and other public venues. Press accounts detail transmission of sound to persons unaware of such use by both developers, along with some description of more disturbing sound exposure, which can include pain even with ear plugs decreasing the noise. **Error! Bookmark not defined.** A non-lethal weapons program director confirms the lack of sound perception by other people nearby on ultrasound voice transmission. Though ultrasound can pass through walls, the encoded sound from ultrasound speakers reflects audibly upon striking hard flat surfaces.

Remote Behavioral Influence Technology Evidence
John J. McMurtrey, M. S., Copyright 2003, 23 Dec. 2003

People discerning remote manipulation by technology capable of such influence have formed protest organizations across the world.

Educated society is uninformed regarding authentic documentation of the development and existence of these technologies, and unaware of the dangers. Complaint of 'hearing voices' and perception of other remote manipulation must receive appropriate scientific and legal investigation with protection. Professional awareness is virtually absent with eminent texts and opinion being presumptuous, without appraisal of the evidence.

Herein is substantiated:

Human wireless internal voice transmission and tracking technologies.

Reports of electroencephalographic (EEG) thought reading capacity, evidence of covert development, and remote EEG capture technology.

References to the use of these, or similar technologies against humans.

NOTE: Financial contributions for John J. McMurtrey's work were made possible by members of Christian Against Mental Slavery, and specifically through the dedication effort of fellow member John Allman, Christians Against Mental Slavery, United Kingdom.

CHAPTER FIVE

"With its grace and carelessness, it seemed to annihilate a whole culture, a whole system of thought, as though all could be swept into nothingness by a single splendid movement of the arm."

- **George Orwell (1984)**

NOTE: Many thanks to Jesse Ventura and the Conspiracy Theory show on TruTV, on December 17, 2012, entitled Brain Invaders for bringing awareness to the capability of technology deployed from what were once the Ground Wave Emergency Network of communication towers and the plight of TIs to mainstream America. An archived show can be accessed online.

Psychological Electronic (psychotronic) technology once termed mind control technology today is highly perfected after many, many, years in research, testing and development programs which date as far back as the 1700s, 1800s and early 1900s on the record. Through the scientific advancements of prolific inventors such as Tesla, Hertz, Maxwell, Hess, and numerous others as the foundation, electromagnetic technology today, in many forms, is nothing less than brilliant. The scientific ideation that it took to get these advancements to their current state of perfection, I must admit myself is amazing.

I personally became aware of the painful reality of advanced psychological manipulation and influence technology, to include physical electronic harassment weaponry, after I became a target, officially called a Targeted Individual (TI) in a growing Targeted Individual community which today numbers in the thousands. Sadly, one thing I would also learn is that credibility continues to be the most powerful obstacle facing a target which works to the advantage of those monitoring the individual. As for me, my experiences of personal pain are documented in "You Are NOT My Big Brother" released in June of 2012.

After publishing You Are Not My Big Brother, this book, Remote Brain Targeting was initially put on hold not wanting the two books to overlap in information or content. However, in February of 2013, after a Targeted Individual contacted me from Tennessee, I live in California, needing specific documented information from this book, alone, to substantiate her claim of targeting in court proceedings, then also another from Northern California requesting anything useful for doctors, I decided this book might be helpful resource as a book compiled of historic information written by others, to enhance legitimacy, and helpful or useful as such, in the ongoing struggle for the credibility TIs face.

It is a well-known fact that in the Targeted Individual community, folks are immediately, without exception, misdiagnosed as delusional, psychotic, or paranoid schizophrenic, when connecting the dots of their personal experiences of remote brain targeting also known as Remote Neural Monitoring. No one wants to officially acknowledge the existence of the technology or legally sanctioned testing, much less the possibility of the existence of the programs written about here, or more importantly, the capability of technology to be secretively deployed against a TI electromagnetically which is unseen or detectable. In my case, there was no help for me forthcoming when reality set in. As a result, I had to help myself by doing the research not only for my credibility but more importantly for my sanity.

The great difficulty in proving the technology exist plays a major role in a TI's discrediting and is strategically useful to those working in these technology testing programs. The fact that many of the technologies are documented by official patents, at the United States Patent and Trademark Office appears meaningless and worthless to some as a source of credibility. The fact that the technology can be deployed subliminally is more frightening and heinous than the average person wants to grasp as a reality or even accept. It is easiest not to think of evil, literally in high places and see the world through rose colored glasses.

Nationwide cell / microwave towers are everywhere and serve as an effective means to disseminate Electromagnetic Low Frequency (ELF) microwave radio waves for various purposes. This includes deployment of the Directed Energy Weapon system in place since the 1980's through the Strategic Defense Initiative (SDI) in the United States. Today all cell phones, televisions, and satellites use microwaves.

Operational control of the primary cyber-security contracts to the federal government involves numerous defense contractors, such the Lockheed Martin Information Systems and Global Solutions. Lockheed's mission and combat support solutions Central Command is located in Norristown, Montgomery County, Pennsylvania. The system is called the Lockheed Martin Government Electronic System (GES) United States Airspace Management – Radar and Satellite Surveillance Automation. However, there are many defense contractors, as cited in You Are Not My Big Brother, playing major and critical roles also contracted by United States government along with ongoing military operation authorized technology testing programs on civilians.

DIRECTED-ENERGY-WEAPON

From Wikipedia, free encyclopedia

A directed-energy weapon (DEW) emits energy in an aimed direction without the means of a projectile. It transfers energy to a target for a desired effect. Intended effects may be non-lethal or lethal.

Some such weapons are real, or are under active research and development.

The energy can come in various forms:

- Electromagnetic radiation, in lasers, masers or heat Particles with mass, in particle beam weapons

- Sound, in sonic weapons. Some such weapons, perhaps most, at present only appear in science fiction, non-functional toys, film props or animation.

In science fiction, these weapons are sometimes known as death rays or ray guns and are usually portrayed as projecting energy at a person or object to kill or destroy. Many modern examples of science fiction have more specific names for directed energy weapons, due to research advances.

These weapons use electromagnetic radiation to deliver heat, mechanical, or electrical energy to a target to cause pain or permanent damage. They can be used against humans, electronic equipment, and military targets. And, depending on the technology when used against equipment, directed electromagnetic energy weapons can operate similarly to omnidirectional electromagnetic pulse (EMP) devices, by inducing destructive voltage within electronic wiring. The difference is that they are directional and can be focused on a specific target using a parabolic reflector. Faraday cages may be used to provide protection from most directed and undirected EMP effects. High-energy radio frequency weapons (HERF) or high-power radio frequency weapons (HPRF) use high intensity radio waves to disrupt electronics.

High Power Microwave devices use microwave radiation, which has a shorter wavelength than radio than directed energy weapons. This equipment, based on microwave, infrasound, neuron-science, biofeedback, and other technology, has the capability to administer a variety of effects when remotely directed at the victim either while inside their homes or away from their homes such as headaches, laser-like burns, rashes, sharp pains that feel like electrical shocks, mood alteration, and neurological trauma. Used against humans, generally is considered 'non-lethal.' However, electromagnetic weaponry, by capability, 'does' pose health threats to humans. In fact, "non-lethal" weapons can be deadly." Some common bio-effects of electromagnetic or other non-lethal weapons include effects to the human central nervous system resulting in physical pain, difficulty breathing, vertigo, nausea, disorientation, or other systemic discomfort, as weapons not directly considered lethal can indeed cause cumulative damage to the human body. These weapons come in the form of vibrations, which can be used to target a specific organ in the body causing scar tissue to form leading to the death of that organ. These weapons can cause heart attacks and strokes. There are approximately 100 plus vibrations that attack the heart. There are vibrations that cause panic attacks and sleep deprivation. There are vibrations which cause a targeted person to go mad and/or crazy, as well as many other capabilities. All that would show up in an autopsy is that the person died from natural causes.

Pulsed microwave manipulation of brain-controlled physiological functioning has been demonstrated in the Arizona State University laboratory of neuroscientist Dr. William J. Tyler, whose work is being funded by the U.S. Army Research Laboratory, according to published articles.

Tyler describes the process as "ultrasonic neuromodulation."

https://blombladivinden.wordpress.com/2011/11/03/us-silently-tortures-americans-with-cell-tower-microwave-weapon/

A recent Popular Science article cites a quotation from Tyler published on the Department of Defense "Armed with Science" blog:

"...(M)y laboratory has engineered a novel technology which implements transcranial pulsed ultrasound to remotely and directly stimulate brain circuits without requiring surgery."

Tyler told Wired.com's "Danger Room:"

"The brain serves all the functions of your body, and if you know the neuroanatomy, then you can start to regulate each one of those functions."

Because the microwave attack system can be precision-targeted to triangulate its multiple beams on unique individuals, this silent torture and impairment technology can and is being used by federal and local law enforcement to impose a regimen of extra-legal electromagnetic incarceration -- a "microwave ghetto" imposed upon citizens without benefit of due process under the law.

According to victim accounts, "innocent but targeted" persons may be subjected to heightened levels of silent torture or impairment if they dare to venture beyond their immediate neighborhoods -- to go "beyond the pale." Operators of the microwave attack weapon obtain the necessary targeting location coordinates from hidden GPS devices; the victim's cellphone; or infrared laser targeting devices trained upon the victim by law enforcement or "community stalkers" who may be affiliated with town watch or community policing groups.

The planning for this "microwave ghetto" is documented in several federally funded studies, such as a March 2001 report from the National Criminal Justice Reference Service.

http://www.ncjrs.gov/pdffiles1/nij/grants/187101.pdf

Medical experts have confirmed that irradiation with microwaves and other radio frequencies can induce injury, illness and disease, from strokes and aneurysms to cataracts and cancer -- and death.

The military aptly applies the descriptor "slow kill" to these weapons systems, capable of targeting and delivering a wide array of electromagnetic microwave and other radio frequency energy with extreme precision -- not "faster than a speeding bullet," in the parlance of "Superman," but at the speed of LIGHT.

Continue reading at NowPublic.com:

U.S. SILENTLY TORTURES AMERICANS WITH CELL TOWER MICROWAVE WEAPON JOURNALIST REPORTS http://nowfact.com/u-s-silently-tortures-americans-with-cell-tower-microwave-weapon/

Patents for microwave and radio frequency directed energy weapons capable of deployment through cellular/microwave

/Gwen communication towers

TITLE:
High Energy Microwave Defense System
United States Patent 4456912

ABSTRACT:

A defense system wherein a radar intercepts and follows a target and includes an antenna array and a low level source of microwave energy for exciting the elements of the array to produce radar returns and define the path. A microwave storage reservoir is supplied upon demand with high energy microwave energy and is coupled to microwave feed channels leading to the elements of the antenna array. Microwave dump switches at the juncture of the channels and the reservoir control flow of energy from the reservoir. Control means responsive at least in part to the presence of returns periodically actuate the dump switches for flow of microwave energy from the reservoir to the target via the antenna array.

INVENTOR:

Ensley, Donald L. (Danville, CA)

Application Number:

05/240420

Publication Date:

06/26/1984

CLAIMS:

What is claimed is:

1. A high energy microwave system for directing energy bursts of microwave radiation at a target comprising:

    a) a microwave cavity having superconducting walls for storing microwave energy,

    b) generating means for feeding waves of microwave energy into said cavity,

    c) transmitting means coupled to said cavity for directing microwave energy toward said target, and

    d) tracking means connected to said transmitting means for initiating flow of microwave energy to said cavity upon locating said target and triggering said transmitting means a predetermined time after initiating said flow to said cavity for directing a burst of electromagnetic energy to said target.

2. A defense system comprising:

    a) a phased array radar adapted to intercept and follow a target path including an antenna array and a low level channel of microwave energy for exciting the elements of said array to produce radar returns and define said path,

    b) a microwave storage reservoir,

c) means to supply upon interception of said target a high level flow of microwave energy to said reservoir,

d) microwave feed channels leading from said reservoir to the elements of said antenna array,

e) microwave dump switches at the juncture of said channels with said reservoir, and

f) means responsive at least in part to said returns for actuating said dump switches for flow of bursts of microwave energy from said reservoir to said target via said antenna array.

3. The combination set forth in claim 2 wherein said reservoir is maintained at a temperature near 0° K. to prevent losses in the walls thereof.

4. The combination set forth in claim 2 wherein said dump switches comprises:

   a) wave guides extending to the wall of said reservoir,

   b) a superconductive layer lining said reservoir and covering each of said wave guides,

   c) a heat conductive dielectric member in each said wave guide backing said layer, and

   d) means to heat said member locally to change the conductivity of said layer at the mouth of each wave guide.

5. The combination set forth in claim 4 wherein said layer is of a niobium material and said dielectric member is quartz.

6. The combination set forth in claim 4 wherein said means to heat said dielectric member is a laser producing a beam directed onto said member.

7. A method of employing high level microwave energy bursts to do work on a moving target at a distance comprising:

a) locating and tracking said target with low energy microwave radar transmissions,

b) upon locating said target initiating flow of high level microwaves to cryogenically controlled storage, and

c) after reaching a predetermined level of energy in storage abruptly dumping said energy via the same path as said radar transmissions to impact said target by locally heating exit locations in the walls of said storage to trigger dumping of said energy.

8. A defense method comprising:

   d) operating a phased array radar to intercept and follow a target path by focusing an antenna array in a low level channel of microwave energy to produce radar returns defining said path,

   e) supplying upon interception of a target a high level flow of microwave energy to a cryogenic microwave storage reservoir, and

   f) responsive at least in part to said returns, actuating localized zones in the walls of said reservoir to render them transmissive for dumping microwave energy from said reservoir onto said target through said antenna array.

9. A high energy microwave system for directing energy bursts of microwave radiation at a target comprising:

   a) a microwave reservoir of the order of 10 meters radius lined with a niobium layer,

   b) a liquid helium system to maintain said layer superconducting during storage of microwave energy in said reservoir,

   c) generating means for feeding microwave energy into said reservoir,

   d) wave guides leading from said reservoir,

e) a phased array radar means coupled to said wave guides for directing microwave energy from said reservoir toward said target,

f) a quartz barrier closure for each of said wave guides contacting said barrier in heat transfer relation, and

g) laser means connected to said phased array radar means for heating said quartz to cause flow of microwave energy from said reservoir in predetermined relation to flow of energy to said reservoir.

DESCRIPTION:

This invention relates to the utilization of high energy electromagnetic waves, and more particularly to a method and system for accumulating and releasing high energy electromagnetic waves in a medium such as the earth's atmosphere.

In national defense, expenditures of time and money are being made in the area of ballistic missiles defenses. One purpose is to provide a defense system specifically directed to destruction of enemy missiles during flight long before a target is reached. The high cost of such systems is occasioned by the necessity of propelling and guiding a physical mass along a path at such speeds as to enable it to collide with or otherwise destroy an enemy missile in flight.

The present invention avoids much of the difficulty encountered in such systems by generating and focusing electromagnetic energy of high intensity to a target to impact the target with electromagnetic energy traveling at the speed of light. Radar systems heretofore have been developed for locating bodies traveling at high speeds through the earth's atmosphere and tracking them. Radars are highly developed and provide reliable information which in an early warning sense heralds the approach of missiles, for example, and provides data to define the path and identify the target point.

The present invention is provided to operate in conjunction with a radar system to generate and store a high quantity of electromagnetic energy and then to dump such energy from storage, feeding the same to an antenna for transmission along the path controlled by the radar. The electromagnetic energy will be focused onto the airborne body with such magnitude as to impart destructive action thereto.

More particularly, in accordance with the present invention, a system is provided for operation of a phased array radar to intercept and follow the path of a body spaced from the radar and includes an antenna array and a source of microwave energy for exciting the array to produce target dependent radar returns.

Further, one or more evacuated microwave storage reservoirs are provided with microwave feed channels leading from the reservoir to elements of the antenna array. Microwave dump switches located at the juncture of the channels and the reservoir permit extraction of energy from the reservoir. Means responsive at least in part to the presence of radar returns actuate the dump switches for flow of microwave energy from the reservoir to the body via the antenna array.

In a further aspect, the invention employs the method of focusing a phased array radar onto a body traveling along a path spaced from the radar, repeatedly storing over substantial intervals microwave energy in a microwave storage reservoir, periodically dumping the energy stored from the radar via channels leading to the array for transmission of high energy microwaves to said body. Microwave energy is fed over a substantial period of time into a reservoir having superconductive walls with windows therein normally coated with a superconductive material which responsive to local heating momentarily become transmissive to the microwave energy stored in the reservoir.

FOR COMPLETE PATENT AND FURTHER DETAILS SEE:

HIGH ENERGY DEFENSE SYSTEM PATENT

http://www.freepatentsonline.com/4456912.html

TITLE:
MULTIFUNCTIONAL RADIO FREQUENCY DIRECTED ENERGY SYSTEM - United States Patent 7629918

ABSTRACT:

An RFDE system includes an RFDE transmitter and at least one RFDE antenna. The RFDE transmitter and antenna direct high power electromagnetic energy towards a target sufficient to cause high energy damage or disruption of the target. The RFDE system further includes a targeting system for locating the target. The targeting system includes a radar transmitter and at least one radar antenna for transmitting and receiving electromagnetic energy to locate the target. The RFDE system also includes an antenna pointing system for aiming the at least one RFDE antenna at the target based on the location of the target as ascertained by the targeting system. Moreover, at least a portion of the radar transmitter or the at least one radar antenna is integrated within at least a portion of the RFDE transmitter or the at least one RFDE antenna.

INVENTORS:

Brown, Kenneth W. (Yucaipa, CA, US)
Canich, David J. (Upland, CA, US)
Berg, Russell F. (Upland, CA, US)

Application Number:

11/300876

Publication Date:

12/08/2009

Filing Date:

12/15/2005

CLAIMS:

What is claimed is:

1. A multi-functional radio frequency directed energy (RFDE) system, comprising: an RFDE transmitter and at least one RFDE antenna for directing high power electromagnetic energy towards a target sufficient to cause high energy damage or disruption of the target; a targeting system for locating the target, the targeting system including a radar transmitter and at least one radar antenna for transmitting and receiving electromagnetic energy to locate the target; and an antenna pointing system for aiming the at least one RFDE antenna at the target based on the location of the target as ascertained by the targeting system, wherein at least a portion of the radar transmitter or the at least one radar antenna is integrated within at least a portion of the RFDE transmitter or the at least one RFDE antenna to provide simultaneous transmission of the high power electromagnetic energy and the electromagnetic energy to locate the target.

2. The multi-functional RFDE system of claim 1, wherein the at least one radar antenna is embodied at least partially within the at least one RFDE antenna.

3. The multi-functional RFDE system of claim 1, wherein the radar transmitter is embodied at least partially within the RFDE transmitter.

4. The multi-functional RFDE system of claim 3, wherein the radar transmitter and the RFDE transmitter comprise a common RF power amplifier.

5. The multi-functional RFDE system of claim 4, wherein the electromagnetic energy for locating the target is at a first frequency, and the high power electromagnetic energy is at a second frequency different from the first.

6. The multi-functional RFDE system of claim 1, wherein the at least one radar antenna functions to transmit the electromagnetic energy for locating the target, and the at least one radar antenna is embodied at least partially in the at least one RFDE antenna.

7. The multi-functional RFDE system of claim 6, wherein the at least one radar antenna includes a first radar antenna that functions to transmit the electromagnetic energy for locating the target and to transmit the high power electromagnetic energy, and a second radar antenna that functions to receive the electromagnetic energy reflected from the target in order to locate the target.

8. The multi-functional RFDE system of claim 7, wherein the first radar antenna comprises a multi-element phased array.

9. The multi-functional RFDE system of claim 6, wherein the at least one radar antenna that functions to transmit the electromagnetic energy for locating the target also functions to receive the electromagnetic energy reflected from the target in order to locate the target.

10. The multi-functional RFDE system of claim 9, wherein the at least one radar antenna comprises a dual-polarized antenna.

11. The multi-functional RFDE system of claim 1, wherein the system comprises a beam combiner for combining the high power electromagnetic energy with the electromagnetic energy for locating the target in a path between the RFDE transmitter and the RFDE antenna.

12. The multi-functional RFDE system of claim 1, wherein the system is configured for operation in a mobile vehicle.

13. The multi-functional RFDE system of claim 12, wherein the mobile vehicle is a wheeled-vehicle.

14. The multi-functional RFDE system of claim 12, wherein the mobile vehicle is an aircraft.

15. A method of operating a multi-functional radio frequency directed energy (RFDE) system, comprising the steps of: utilizing an RFDE transmitter and at least one RFDE antenna to direct high power electromagnetic energy towards a target sufficient to cause high energy damage or disruption of the target; utilizing a targeting system to locate the target, the targeting system including a radar transmitter and at least one radar antenna for transmitting and receiving electromagnetic energy to locate the target; aiming the at least one RFDE antenna at the target based on the location of the target as ascertained by the targeting system; and integrating at least a portion of the radar transmitter or the at least one radar antenna within at least a portion of the RFDE transmitter or the at least one RFDE antenna to provide simultaneous transmission of the high power electromagnetic energy and the electromagnetic energy to locate the target.

16. The method of claim 15, wherein the at least one radar antenna is embodied at least partially within the at least one RFDE antenna.

17. The method of claim 15, wherein the radar transmitter is embodied at least partially within the RFDE transmitter.

18. The method of claim 17, wherein the radar transmitter and the RFDE transmitter comprise a common RF power amplifier.

19. The method of claim 18, wherein the electromagnetic energy for locating the target is at a first frequency, and the high power electromagnetic energy is at a second frequency different from the first.

20. The method of claim 15, wherein the at least one radar antenna functions to transmit the electromagnetic energy for locating the target, and the at least one radar antenna is embodied at least partially in the at least one RFDE antenna.

21. A multi-functional radio frequency directed energy (RFDE) system, comprising: an RFDE transmitter and at least one RFDE antenna for directing high power electromagnetic energy towards a target sufficient to cause high energy damage or disruption of the target; a targeting system for locating the target, the targeting system including a radar transmitter and at least one radar antenna for transmitting and receiving electromagnetic energy to locate the target; and an antenna pointing system for aiming the at least one RFDE antenna at the target based on the location of the target as ascertained by the targeting system, wherein at least a portion of the radar transmitter or the at least one radar antenna is integrated within at least a portion of the RFDE transmitter or the at least one RFDE antenna and the high power electromagnetic energy is used as the electromagnetic energy to locate the target.

22. A method of operating a multi-functional radio frequency directed energy (RFDE) system, comprising the steps of: utilizing an RFDE transmitter and at least one RFDE antenna to direct high power electromagnetic energy towards a target sufficient to cause high energy damage or disruption of the target; utilizing a targeting system to locate the target, the targeting system including a radar transmitter and at least one radar antenna for transmitting and receiving electromagnetic energy to locate the target; aiming the at least one RFDE antenna at the target based on the location of the target as ascertained by the targeting system; and integrating at least a portion of the radar transmitter or the at least one radar antenna within at least a portion of the RFDE transmitter or the at least one RFDE antenna, wherein the high power electromagnetic energy is used as the electromagnetic energy to locate the target.

DESCRIPTION:

TECHNICAL FIELD

The present invention relates generally to radio frequency directed energy (RFDE) systems, and more particularly to multifunctional type RFDE systems.

BACKGROUND OF THE INVENTION

Radio frequency directed energy (RFDE) systems are known in the art for directing high power RF, microwave and/or millimeter wave electromagnetic energy to destroy or disrupt a target. Although RFDE systems typically serve as military weapons, RFDE systems need not be limited to weapon systems. For example, RFDE systems of the present invention may be used for non-military purposes such as destroying or disrupting foreign objects, contaminants, undesirable atmospheric conditions, or other types of targets.

As for weapon systems, it is important to distinguish between an RFDE weapon system and an electronic warfare system. A primary difference between an RFDE weapon and an electronic warfare system is power and kill mode. An electronic warfare system makes use of a priori knowledge of a target it is designed to jam or disrupt. An electronic warfare system uses such a priori knowledge of a target's characteristics (e.g., frequency of operation, method of operation, etc.) to disrupt or confuse the target with "finesse", or a relatively low amount of power.

On the other hand, an RFDE weapon system can go after a broad range of targets (electronics, biological, ordinance, structures, etc.) due to its relatively large radiated power. A priori knowledge of the intended target characteristics is typically not required because the RFDE weapon either burns-out or overwhelms its target by the shear amount of power it radiates.

An ongoing problem with RFDE systems is targeting—accurately pointing the RF directed energy beam at the intended target and establishing an accurate range from the system to the target. To date, the RFDE system targeting problem has been addressed by using what may be referred to as auxiliary add-on systems. These add-on systems could include a stand-alone radar system, a stand-alone laser range finder, stand-alone optical or infrared imaging system, etc. However, these add-on systems add significant cost to the RFDE system. In addition, these add-on systems add significant complexity by requiring calibration of the alignment between the RFDE system and the stand-alone targeting system.

FOR COMPLETE PATENT AND FURTHER DETAILS SEE:

http://www.freepatentsonline.com/7629918.html.

CHAPTER SIX

"In philosophy, or religion, or ethics, or politics, two and two might make five, but when one was designing a gun or an aero plane they had to make four."

- George Orwell (1984)

A multimillion-dollar program of research on germ warfare and on methods to alter or control human memory and behavior through the use of drugs, electricity, sensory deprivation, hypnosis, and other means spanned the years from, as a focal point from 1940 onward. This research involved 185 researchers at 88 non-governmental institutions, including 44 colleges and universities. The magnitude, projects, scope and duration seemed to justify the conclusion of former State Department officer John Marks that "the intelligence community changed the face of the scientific community during the 1950s and early 1960s."

Experiments and research have continued unobstructed for many years. As recent as 2006, researchers at the University of Washington were working on an electronic chip (implant) that could control nerve connections in the area of the brain that controls movement. In 2008, the Army gave the University of California, Irvine, scientist in California a 4 million dollar grant to study Synthetic Telepathy. Synthetic Telepathy is defined on synthetictelepathy.com as "the art of electronically transferring thought directly to and from a brain." The primary objective of the University of California, Irvine's scientific grant as stated below by the University says:

Under construction by contractors with top-secret clearances, is the blandly named Utah Data Center being built for the National Security Agency in Bluffdale, Utah. This project of immense secrecy is the final piece in a complex puzzle assembled over the past decade. Its purpose: to intercept, decipher, analyze, and store vast swaths of the world's communications as they zap down from satellites and zip through the underground and undersea cables of international, foreign, and domestic networks. The heavily fortified $2 billion center should be up and running in September 2013. The facility will house a powerful brain-computer/supercomputer.

Real time imagery is cost-free to a requesting agency today as a result of this system:

IRIDIUM SATELLITE SYSTEM

The Iridium Satellite System is one of 66 low-earth orbiting (LEO) satellites which comprise this system. The original company went bankrupt so the US Department of Defense assisted in making sure the system stayed in operation. The system is currently maintained by the Boeing Company. Iridium multiple spot beams (cells) blanket the total earth surface. The original design uses 77 satellites. This system has transmitting and receiving coverage of the entire planet. Each circle is a coverage sector.

Today, satellite can track your every move by computer generated mapping of your bio-energetic signature body biometrics by constantly scanning an area to find you.

"DISA or Defense Information Systems Network, DISN is a combat support agency responsible for planning, developing, fielding, operating, and supporting command, control communications, and information systems that serve the needs of the president, Vice President, the Secretary of Defense, the Joint Chiefs of Staff and

Combatant Commanders and the other Department of Defense (DOD) components, under all conditions of peace and war." This is how it is stated on the actual website. Their Mission Statement as stated also: "To provide information and services that is mission critical to the operation of the worldwide IP router of the Defense Information Systems Network and other DOD sponsored networks."

As I continued to read, the site it further stated that satellite communications are one of the many services, provided saying:

"Satellite communication (SATCOM) expands the reach of the network and delivers communication capabilities anytime, anywhere, and in any environment. Within the DOD, DISA leads the effort to provide innovative, responsive, and cost effective SATCOM services to our customers. DISA is involved in all segments of the SATCOM environment including commercial SATCOM. They are the sole authorized provider of commercial SATCOM services to the DOD."

The law allowing military assistance, such as SATCOM in the civilian arena is stated in the United States Code below which says:

Inside the United States (as well as abroad), DOD support for law enforcement agencies is authorized in accordance with Chapter 18 of Title 10 of the U.S. Code. The legislation contains both explicit grants of authority and restrictions on the use of that authority for DOD assistance to law enforcement agencies—federal, state, and local—particularly in the form of information and equipment. Section 371 specifically authorizes the Secretary of Defense to share information acquired during military operations, and encourages the armed forces to plan their activities with an eye to the production of incidental civilian benefits. Under sections 372 through 374, DOD equipment and facilities, including intelligence collection assets, may be made available to civilian authorities.

Signals intelligence collection of targets within the United States is governed by the Foreign Intelligence Surveillance Act (FISA.) The Foreign Intelligence Surveillance Act of 1978 ("FISA" Pub.L. 95-511, 92 Stat. 1783, enacted October 25, 1978, 50 U.S.C. ch.36, S. 1566) is an Act of Congress which prescribes procedures for the physical and electronic surveillance and collection of "foreign intelligence information" between "foreign powers" and "agents of foreign powers" (which may include American citizens and permanent residents suspected of being engaged in espionage and violating U.S. law on territory under United States control) and domestically the Electronic Communication Privacy Act (ECPA.)

The Electronic Communications Privacy Act of 1986 (ECPA Pub. Law 99-508, Oct. 21, 1986, 100 Stat. 1848, 18 U.S.C. § 2510) was enacted by the United States Congress to extend government restrictions on wire taps from telephone calls to include transmissions of electronic data by computer. Specifically, ECPA was an amendment to Title III of the Omnibus Crime Control and Safe Streets Act of 1968 (the Wiretap Statute), which was primarily designed to prevent unauthorized government access to private electronic communications.

The ECPA also added new provisions prohibiting access to stored electronic communications, i.e., the Stored Communications Act, 18 U.S.C. §§ 2701-12. The ECPA also included so-called pen/trap provisions that permit the tracing of telephone communications. §§ 3121-27. Later, the ECPA was amended, and weakened to some extent, by some provisions of the USA PATRIOT Act. In addition, Section 2709 of the Act, which allowed the FBI to issue National Security Letters (NSLs) to Internet service providers (ISPs) ordering them to disclose records about their customers, was ruled unconstitutional under the First (and possibly Fourth) Amendments in ACLU v. Ashcroft (2004). It is thought that this could be applied to other uses of National Security Letters.

EXECUTIVE ORDER 12333

E.O. 12333 augments statutory intelligence authority for the Secretary of Defense as well as relevant offices and agencies within the Department. The functions of the National Reconnaissance Office (NRO) are described in paragraph 1.7(e), and include the production and dissemination of geospatial intelligence information and data "for foreign intelligence and counterintelligence purposes to support national and departmental missions," as well as the provision of "geospatial intelligence support for national and departmental requirements and for the conduct of military operations." Assistance to law enforcement agencies is covered in paragraph 2.6 of E.O. 12333, which authorizes agencies within the Intelligence Community to participate in law enforcement activities to investigate or prevent clandestine intelligence activities, international terrorist activities, or narcotics trafficking activities. The order also permits the intelligence elements to provide specialized equipment, technical knowledge, or assistance of expert personnel for use by any department or agency, or, when lives are endangered, to support local law enforcement agencies.

E.O. 12333 requires agencies within the Intelligence Community to use "the least intrusive collection techniques feasible within the United States or directed against United States persons abroad." Monitoring devices may be used only "in accordance with procedures established by the head of the agency concerned and approved by the Attorney General. Such procedures shall protect constitutional and other legal rights and limit use of such information to lawful governmental purposes." The Attorney General is delegated the authority to approve the use, within the United States or against a United States person abroad, of "any technique for which a warrant would be required if undertaken for law enforcement purposes, provided that such techniques shall not be undertaken unless the Attorney General has determined in each case that there is probable cause to believe that the technique is directed against a foreign power or an agent of a foreign power."

Electronic Surveillance by definition, Title 50, Chapter 36, Subchapter 1, § 1801 excerpt defining Electronic surveillance by law:

(f) "Electronic surveillance" means—

the acquisition by an electronic, mechanical, or other surveillance device of the contents of any wire or radio communication sent by or intended to be received by a particular, known United States person who is in the United States, if the contents are acquired by intentionally targeting that United States person, under circumstances in which a person has a reasonable expectation of privacy and a warrant would be required for law enforcement purposes:

the acquisition by an electronic, mechanical, or other surveillance device of the contents of any wire communication to or from a person in the United States, without the consent of any party thereto, if such acquisition occurs in the United States, but does not include the acquisition of those communications of computer trespassers that would be permissible under section 2511(2)(i) of title 18;

the intentional acquisition by an electronic, mechanical, or other surveillance device of the contents of any radio communication, under circumstances in which a person has a reasonable expectation of privacy and a warrant would be required for law enforcement purposes, and if both the sender and all intended recipients are located within the United States; or,

the installation or use of an electronic, mechanical, or other surveillance device in the United States for monitoring to acquire information, other than from a wire or radio communication, under circumstances in which a person has a reasonable expectation of privacy and a warrant would be required for law enforcement purposes.

Title 50, Chapter 36, Subchapter I, § 1802

§ 1802. Electronic surveillance authorization without court order; certification by Attorney General; reports to Congressional committees; transmittal under seal; duties and compensation of communication common carrier; applications; jurisdiction of court.

Hand in Hand with this type of Electronic Surveillance involves the community, neighbors, employers, businesses and even family and friends in what is called Gang stalking. In most cases, from my experience, law enforcement, are not about telling people outrageous things about the "Target" to enlist the communities support. If effective, the "Target's" life become literally a living hell as strangers appear to have resentments or even hate for the person with is often unexplainable for the person being Targeted. In fact, neighbors, in adjoining apartments are provided with an array of technology, to assist the effort to victimize a person, such as computers surveillance software and it appears smaller versions or hand held equipment microwave technology or technology allowing through the wall observance. Although it is illegal, according to the Title below:

Title 18 Part I Chapter 119 2511

"Interception and disclosure of wire, oral, or electronic communications prohibited an offense under this subsection that consists of or relates to the interception of a satellite transmission that is not encrypted or scrambled and that is transmitted:

1) Except as otherwise specifically provided in this chapter any person who:

a) intentionally intercepts, endeavors to intercept, or procures any other person to intercept or endeavor to intercept, any wire, oral, or electronic communication;

b) intentionally uses, endeavors to use, or procures any other person to use or endeavor to use any electronic, mechanical, or other device to intercept any oral communication when

    i. such device is affixed to, or otherwise transmits a signal through, a wire, cable, or other like connection used in wire communication (i) intentionally discloses, or endeavors to disclose, to any other person the contents of any wire, oral, or electronic communication, intercepted by means authorized by sections 2511 (2)(a)(ii), 2511 (2)(b)–(c), 2511(2)(e), 2516, and 2518 of this chapter, shall be punished as provided in subsection (4) or shall be subject to suit as provided in subsection (5). (4) (a) except as provided in paragraph (b) of this subsection or in subsection (5), whoever violates subsection (1) of this section shall be fined under this title or imprisoned not more than five years, or both. As you can see this is the penalty for such offends.

TITLE 50--WAR AND NATIONAL DEFENSE

CHAPTER 32--CHEMICAL AND BIOLOGICAL

WARFARE PROGRAM

Sec. 1520a. Restrictions on use of human subjects for testing of chemical or biological agents

A. Prohibited activities

The Secretary of Defense may not conduct (directly or by contract)

1) any test or experiment involving the use of a chemical agent or biological agent on a civilian population; or

2) any other testing of a chemical agent or biological agent on human subjects.

B. Exceptions

Subject to subsections (c), (d), and (e) of this section, the prohibition in subsection (a) of this section does not apply to a test or experiment carried out for any of the following purposes:

1) Any peaceful purpose that is related to a medical, therapeutic, pharmaceutical, agricultural, industrial, or research activity.

2) Any purpose that is directly related to protection against toxic chemicals or biological weapons and agents.

3) Any law enforcement purpose, including any purpose related to riot control.

C. Informed consent required

1) The Secretary of Defense may conduct a test or experiment described in subsection.

2) of this section only if informed consent to the testing was obtained from each human subject in advance of the testing on that subject.

D. Prior notice to Congress

Not later than 30 days after the date of final approval within the Department of Defense of plans for any experiment or study to be conducted by the Department of Defense (whether directly or under contract) involving the use of human subjects for the testing of a chemical agent or a biological agent, the Secretary of Defense shall submit to the Committee on Armed Services of the Senate and the Committee on Armed Services of the House of Representatives a report Setting, forth a full accounting of those plans, and the experiment or study may then be conducted only after the end of the 30-day period beginning on the date such report is received by those committees.

E. "Biological agent'" defined

In this section, the term ``biological agent" means any microorganism (including bacteria, viruses, fungi, rickettsia, or protozoa), pathogen, or infectious substance, and any naturally occurring, bioengineered, or synthesized component of any such microorganism, pathogen, or infectious substance, whatever its origin or method of production, that is capable of causing:

1) death, disease, or other biological malfunction in a human, an animal, a plant, or another living organism;

2) deterioration of food, water, equipment, supplies, or materials of any kind; or

3) deleterious alteration of the environment.

(Public Law 105-85, Div. A, title X, Sec. 1078, Nov. 18, 1997, 111 Stat. 1915; Public. Law 106-65, div. A, title X, Sec. 1067(4), Oct. 5, 1999, 113 Stat. 774.)

Codification

Section is comprised of section 1078 of Pub. L. 105-85. Subsec. (f) of section 1078 of Pub. L. 105-85 amended section 1523

(b) of this title. Subsec.

(g) of section 1078 of Pub. L. 105-85 repealed section 1520 of this title.

Section was enacted as part of the National Defense Authorization Act for Fiscal Year 1998, and not as part of Pub. L. 91-121, title IV, Sec. 409, Nov. 19, 1969, 83 Stat. 209, which comprises this chapter.

Amendments

1999--Subsec. (d). Pub. L. 106-65 substituted ``and the Committee on Armed Services" for ``and the Committee on National Security".

THE PATRIOT ACT

Commonly known as the "Patriot Act") is an Act of the U.S. Congress that was signed into law by President George W. Bush on October 26, 2001. The title of the Act is a contrived three letter initialism (USA) preceding a seven letter acronym (PATRIOT), which in combination stand for Uniting and Strengthening America by Providing Appropriate Tools Required to Intercept and Obstruct Terrorism Act of 2001. The Act dramatically reduced restrictions on law enforcement agencies' ability to search telephone, e-mail communications, medical, financial, and other records; eased restrictions on foreign intelligence gathering within the United States; expanded the Secretary of the Treasury's authority to regulate financial transactions, particularly those involving foreign individuals and entities; and broadened the discretion of law enforcement and immigration authorities in detaining and deporting immigrants suspected of terrorism-related acts. The act also expanded the definition of terrorism to include domestic terrorism, thus enlarging the number of activities to which the USA PATRIOT Act's expanded law enforcement powers can be applied.

Just before the midnight deadline on May 26, 2011, President Barack Obama signed a 4-year extension of three key provisions in the USA Patriot Act: roving wiretaps, searches of business records (the "library records provision"), and conducting surveillance of "lone wolves" — individuals suspected of terrorist-related activities not linked to terrorist groups.

Details

The Act was passed in the House by 357 to 66 (of 435) and in the Senate by 98 to 1 and was supported by members of both the Republican and Democratic parties.

Opponents of the law have criticized its authorization of indefinite detentions of immigrants; searches through which law enforcement officers search a home or business without the owner's or the occupant's permission or knowledge; the expanded use of National Security Letters, which allows the Federal Bureau of Investigation (FBI) to search telephone, e-mail, and financial records without a court order, and the expanded access of law enforcement agencies to business records, including library and financial records. Since its passage, several legal challenges have been brought against the act, and Federal courts have ruled that a number of provisions are unconstitutional.

CHAPTER SEVEN

"He who controls the past controls the future. He who controls the present controls the past."

- George Orwell (1984)

Various excerpts from Wikipedia

Nikola Tesla, a physicist and inventor, came to the United States, arriving in New York on April 7, 1882. He was age 25 coming from Antwerp. Though Tesla immigrated as a humble laborer, this title hardly befitted the man who would become one of the most prolific inventors in history if not one of the greatest. Tesla was an expert mechanical and electrical engineer and ultimately a very important contributor to the birth of commercial electricity. With close to 700 patents, though say closer to 1,000 patents to his credit ranging from alternating currents (AC) power, radio, x-ray, radar, solar, the Tesla Shied, TeleForce, and the induction motor. Tesla's research would become the foundation for some of the most brilliant inventions and technological advances in the 19th and 20th and 21st Centuries.

The Warden Clyffe Tower (1901–1917) also known as the Tesla Tower, was an early wireless telecommunications tower designed by Nikola Tesla. It was intended for commercial trans-Atlantic wireless telephone, broadcasting, and to demonstrate the transmission of power without interconnecting wires. The core facility was not completed due to financial problems and was never fully operational. And, there was also no motivation to provide free energy at that time or even today.

Nikola Tesla is known for many revolutionary developments however, it would be his work in the field of electromagnetism that would have the greatest impact on the world of scientific development. Tesla demonstrated wireless energy transmission as early as 1891. The Tesla effect is a term for an application of this type of electrical conduction (that is, the movement of energy through space and matter, not just the production of voltage across a conductor.)

This view changed, however, with the publication of James Clerk Maxwell's 1873 Treatise on Electricity and Magnetism in which the interactions of positive and negative charges were shown. Of all the great inventions and discoveries of Nikola Tesla, nothing stood out more as holding great potential and benefit to humanity positively than his discovery of Radiant Energy in 1889. The series of observations that led to the discovery of Radiant energy initially grew out of experiments that Tesla had conducted in an attempt to duplicate the results that Heinrich Hertz, Heinrich Rudolf Hertz was a German physicist who clarified and expanded the electromagnetic theory of light that had been put forth by Maxwell. (Originally electricity and magnetism were thought of as two separate forces. to be regulated by one force.) Hertz was the first to satisfactorily demonstrate the existence of electromagnetic waves by building an apparatus to produce and detect VHF or UHF radio waves affirming the existence of electromagnetic waves. Hertz announced the discovery in 1887.

From 1893 to 1895, Tesla was the first to transmit electromagnetic energy without wires building the first radio transmitter. In 1897, Tesla research of radiation, led to setting up the basic formulation of cosmic rays. Radiant energy is what Tesla called the energy of electromagnetic waves and Tesla sought to harness this energy. Later in his life, Tesla would say:

In the "Brooklyn Eagle," dated, July 10, 1932:

"I have harnessed the cosmic rays and caused them to operate a motive device. Cosmic ray investigation is a subject that is very close to me. I was the first to discover these rays and I naturally feel toward them as I would toward my own flesh and blood. I have advanced a theory of the cosmic rays and at every step of my investigations

Then in, the "New York American," November 1st, 1933: Device to Harness Cosmic Energy Claimed by Tesla:

"This new power for the driving of the world's machinery will be derived from the energy which operates the universe, the cosmic energy, whose central source for the earth is the sun and which is everywhere present in unlimited quantities."

Tesla was intrigued by being able to harvest the energy prevalent throughout space. He knew that it would be simply a matter of time before science and technology would accomplish attaching machinery to the radiant energy of nature and stating: "Ere many generations pass, our machinery will be driven by a power obtainable at any point of the universe." In 1937, he wrote a treatise entitled "The Art of Projecting Concentrated Non-dispersive Energy through the Natural Media," which centered on charged particle beams and from 1900s until his death worked on what the press called a peace ray, death ray or Teleforce. Today, these weapons, in various forms are called ray guns and stun guns and Directed Energy Weapons, and Active Denial System. The technology was invented for military application by a letter Tesla wrote to J. P. Morgan, Jr., on November 29, 1934, regarding the Teleforce in which it can be seen that the device was intended for use in national defense. Tesla's work on particle beam weapons can be traced prior to, to 1893 with his invention of a button lamp, and again to 1896 when he replicated the work of William Roentgen, discoverer of X-rays.

At about the year 1918, Tesla apparently also had a laser-like apparatus that he shot at the moon. From studying his great 1893 work, in the "The Inventions, Researches, and Writing of Nikola Tesla," it is apparent that the button lamp discussed in the writings had all of the components necessary to create a laser beam or the ruby laser. A laser is a device that emits light (electromagnetic radiation) through a process of optical amplification based on the stimulated emission of photons. So, advanced were many of Tesla's inventions that development continued after his death through secret government programs and while he lived, secrecy shrouded the inventions almost immediately after the discovery itself. This was specifically true regarding alternating current (AC), electromagnetic energy, generators, and coils which later became the foundation for HAARP, radio transmission, energy-saving devices, and wireless transmission.

Tesla's research and development set the stage for technology using electromagnetic microwaves today, or "Star Wars," approved for use by the Strategic Defense Initiative (SDI) of 1983, in Ronald Reagan Administration. Tesla's radar is one medium of delivery by satellites. Tesla experimented with lightning (HAARP), power generating systems, and electrical transmission systems. Wireless radio, which history books assign to Marconi, was actually Tesla's concept, as, again, were RADAR and florescent lighting. Radar is an object detection system which uses electromagnetic waves, specifically radio waves, to determine the range, altitude, direction, or speed of both moving and fixed objects. Although the technology was first demonstrated in the United States in December 1934, it was only when war clouds loomed that United States (U.S.) military authorities recognized the great potential of radio-based detection and tracking, and began the development of ship and land-based systems. The first of these were fielded by the U.S. Navy in early 1940, and a year later by the U.S. Army. The acronym RADAR (Radio Detection and Ranging) was coined by the U.S. Navy in 1940, and the subsequent name "radar" was soon widely used.

A Tesla Coil is a type of resonant transformer circuit invented by Tesla around 1891. Tesla used these coils to conduct innovative experiments in electrical lighting, phosphorescence, x-ray generation, high frequency alternating current phenomena, electrotherapy, and again, the transmission of electrical energy without wires. His Tesla

Coil is perhaps a symbol of one of his major goals: the cheap and safe wireless transmission of electrical power. Tesla's as a scientist was motivated by a humanitarian desire.

The High Frequency Active Auroral Research Program (HAARP) is an ionospheric research program jointly funded by the U.S. Air Force, the U.S. Navy, the University of Alaska and the Defense Advanced Research Projects Agency (DARPA) and is located near Gakona, in Alaska. The program officially describes itself as "A scientific endeavor aimed at studying the properties and behavior of the ionosphere, with particular emphasis on being able to understand and use it to enhance communications and surveillance systems for both civilian and defense purposes." The project is a giant array of microwave towers. It's billion plus watts of power can impact an entire nation or a complete hemisphere of the world, and has a delivery source compatible with Digital television transmission and as a possible source for Electromagnetic or Extra Low Frequency (ELF). The ELF ability thereby allows the capability of an inaudible verbal message carrier system. HAARP, also called the Ionospheric Research Instrument (IRI) is also a high-power transmitter facility used to excite portions of the Ionosphere. More specifically, HAARP is a highly controversial high frequency radio transmitter, or "ionospheric heater".

"The $30 million United States Pentagon project is made to beam more than 1.7 gig watts (billion watts) of radiated power into the ionosphere -- the electrically charged layer above Earth's atmosphere. Put simply, the apparatus is a reversal of a radio telescope -- just transmitting instead of receiving. It will 'boil the upper atmosphere'. After heating and disturbing the ionosphere, the radiations will bounce back onto the earth in the form of long waves which penetrate our bodies, the ground and the oceans" and I found endangering fish life.

Excerpt from: "Angels Don't Play This HAARP" Dr. Nick Belgich, page 8. NOTE: Dr. Belgich is a world-renowned expert in this field.

The GWEN Towers, originally conceived during the early days of the Ronald Reagan Presidential Administration, of 1981 to 1989. The Air Force placed a tentative initial operating capability for GWEN by

January 1992. The Ground Wave Emergency Network (GWEN) was a command and control communications system originally intended for use by the United States government to facilitate military communications before, during, or after a nuclear war. Specifically, GWEN was constructed to survive the effects of a high-altitude nuclear explosion electromagnetic pulse and to insure that the United States President could give a launching order to strategic nuclear bombers. GWEN transmitters have many different functions, including controlling the weather, mind, and it is said the ability to control the behavior and mood of the populace. The tower is documented as also having the capability to send synthetic telepathy disguised as infrasound, to victims of U.S. government mind-control by microchip or Verichip implants. GWEN (Ground Wave Emergency Network) transmitters, placed 200 miles apart across the USA, allow specific frequencies to be tailored to the geomagnetic-field strength in each area, allowing the magnetic field to be altered. The Creator designed living beings to resonate to this natural frequency pulsation in order to evolve harmoniously. Essentially, the tower has the capability to confuse, distort and change this frequency.

Numerous accounts of people declaring microchip implantations have been verified by people in the United States as well as in countries such as Sweden. The brain transmitter has also been used in the Soviet Union. However, due to further advances, of satellite delivered Neurophone or Voice to Skull, and microwave subliminal message carrying technology, implants have been nearly made obsolete. This is largely due to the capability of "Synthetic Telepathy" deliverance which is equally effective and requires no physical manipulation of the subject medically. However, those implanted before these means evolved are still tracked by GWEN Towers. Today, implants are now smaller than a hair's width and can be injected by vaccine and flu shots or put into food. These 'biochips' then circulate in the bloodstream and lodge in the brain, enabling the victims to hear 'voices' via the implant. There are many types of implants currently, and it is estimated that 1 in 40 people are recipients of these tiny implants due to government testing programs disguised as alien abductions. If alien abductions are factual, the sheer number of people claiming the experiences, logically, should initiate and activate state of immediate urgency in a national emergency. Regarding implants, there are those, however, that have suggested that 1 in 20 implantations might be a more accurate statistic for implantation.

The Russians openly market a small version of their weather-engineering system called Elate, which can fine-tune weather patterns over a 200-mile area and have the same range as the GWEN unit. One such system operates at the Moscow airport.

Although Nikola, made great scientific contributions, Tesla sadly died penniless at age 86 o, 1943 suffering from what appeared to be Anorexia and is today called Obsessive Compulsive Disorder.

In, December 1901, Guglielmo Marconi established wireless communication between Britain and the Newfoundland, Canada, earning him the Nobel Prize in 1909. But much of Marconi's work was not original. In 1864, James Maxwell theorized electromagnetic waves. In 1887, Heinrich Hertz proved Maxwell's theories. Later, Sir Oliver Lodge extended the Hertz prototype system. The Brandley coherer increased the distance messages to be transmitted. The coherer was perfected by Marconi. However, the heart of radio transmission is based upon four tuned circuits for transmitting and receiving. This is Tesla's original concept demonstrated in his famous lecture at the Franklin Institute in Philadelphia in 1893. The four circuits, used in two pairs, are still a fundamental part of all radio and television equipment today. The United States Supreme Court, in 1943 held Marconi's most important patent invalid, recognizing Tesla's more significant contribution as the inventor of radio technology. The Nobel Prize given to Marconi was really Nikola Tesla's Nobel Prize.

Tesla's diary contains explanations of his experiments concerning the ionosphere and the ground's telluric currents via transverse waves and longitudinal waves. At his lab, Tesla proved that the earth was a conductor, and he produced artificial lightning (with discharges consisting of millions of volts, and up to 135 feet long.)

The Russian Woodpecker transmitter is the Russian attempt at global mind control and it is a system similar to HAARP. When Nikola Tesla revealed that power could be transmitted through the ground using ELF waves saying, "Nothing stops or weakens these signals" the foundation was laid. The Russians retrieved Tesla 's papers 'officially' when they were returned to Yugoslavia after his death. However, proof that particle beam weapons were given to the Soviets was established by Colonel Tom Bearden. Thomas E. Bearden, Ph.D., Nuclear Engineer, retired Lieutenant Colonel (U.S. Army), CEO of CTEC, Inc., Director of the Association of Distinguished American Scientists, and Fellow Emeritus of the Alpha Foundation's Institute for Advanced Study. He is a theoretical conceptualist active in the study of scalar electromagnetics, advanced electrodynamics, unified field theory, KGB energetics weapons and phenomena, free energy systems, electromagnetic healing via the unified field action of extended Sachs-Evans electrodynamics, and human development. He is particularly known for his work establishing a theory of over unity electrical power systems, scalar electromagnetic weapons, energetics weapons, and the use of time-as-energy in both power systems and the mind-body interaction. In the May 2, 1977 issue of Aviation Week, Colonel Bearden pointed out that the Soviet had successfully gain Tesla's original patent. The article also displays a picture of a Soviet particle beam weapon, (along with the accompanying 7000-word article) that is almost a carbon copy of the picture in Tesla's 1937 patent application. The technology would be able to shoot down incoming planes at distances of about 300 miles.

The fact that Tesla died penniless was not only due to his poor business sense, Tesla was a scientist, not a business man, but also due not only to him receiving little monetary reward for his numerous inventions, but due also to his inventions being smuggled out of the country, to assist in the political military goals of other nations in man's continued quest then and today for military dominance. Tesla himself tried to sell his inventions during World War II to not only the United States, but to England, the Soviet Union and Yugoslavia.

Not long after Tesla's death in 1943, Dr. John G. Trump, National Defense Research Committee (NDRC) of the Office of Scientific Research and Development, Technical Aids, Division 14, went over 150,000 pages of Tesla's documents and notes hoarded in various storage locations and also in the basement of the New Yorker Hotel where Tesla stayed until his death.

The National Defense Research Committee (NDRC) originated as an organization created "to coordinate, supervise, and conduct scientific research in areas underlying the development, production, and use of mechanisms and devices of warfare" in the United States from June 27, 1940 until June 28, 1941. The NDRC was initiated, by Vannevar Bush, an American engineer and Science Administrator and Director, Office of Science Research and Development, which superseded the NDRC. The Office of Research and Development was approved by and reported directly to Franklin Roosevelt. It was created to pull together the resources of the military and civilian scientific technological communities which were divided and skeptical of each other, into one agency for overall government scientific military advancement. As was the case with most research in the scientific military arena, its work was done in the strictest of confidence and secrecy.

The NDRC began research into what would become some of the most important technology during World War II, including radar and the atomic bomb initially. Vinegar Bush, who appears to have no relationship the Bush family of political lineage, was later one of the founders of Raytheon. Raytheon today is still one of the top U.S. government's top contractors in the area of a significant defense, in areas such as energy weapons, and the Active Denial System, to include the stun guns, etc., now used by law enforcement.

After the war the demand for technology, research and advancement necessitated and resulted in Operation Paperclip and ODESSA where some of the first mind control experiments began. Operation Paperclip was the initiative to relocate top German scientist to the United States following World War II and it was during these times that actual testing began in depth in the actual endeavor of physical and psychological manipulation of the human brain for, let's say other than health reasons.

The CIA is the successor of the Office of Strategic Services (OSS) formed during World War II to coordinate espionage activities against the Axis Powers for the branches of the United States Armed Forces. The National Security Act of 1947 established the CIA. On December 31, 1948, the CIA formed the Office of Scientific Intelligence (OSI). The post of Director, Central Intelligence (DCI) was established by President Harry Truman on January 23, 1946. The office existed from January 1946 to April 2005 and was replaced by Director of National Intelligence (DNI) and Director of the Central Intelligence Agency (D/CIA). In 1962, the CIA formed the Deputy Directorate of Research.

The CIA created an Electromagnetic Research (EMR) laboratory at Allan Memorial, a Montreal, Canada, research facility created in 1943 as Tesla's invention moved through government agencies.

Electromagnetic radiation (often abbreviated E-M radiation or EMR) is a form of energy exhibiting wave-like behavior as it travels through space. EMR has both electric and magnetic field components, which oscillate in phase perpendicular to each other and perpendicular to the direction of energy propagation. Electromagnetic radiation is classified according to the frequency of its wave. In the area of increasing frequency and decreasing wavelength, there are radio waves, microwaves, infrared radiation, visible light, ultraviolet radiation, X-rays and gamma rays (see Electromagnetic spectrum). The eyes of various organisms sense a small and somewhat variable window of frequencies called the visible spectrum. The photon is the quantum of the electromagnetic interaction and the basic "unit" of light and all other forms of electromagnetic radiation and is also the force carrier for the electromagnetic force. Electromagnetic radiation carries energy and momentum that may be imparted to matter with which it interacts as is the foundation of today's Star Wars technology using particle beam. Electromagnetic (radiation) weapons are a type of directed energy weapons which use electromagnetic radiation to deliver heat, mechanical, or electrical energy to a 'target' to cause various, sometimes very subtle, effects. They can be used against humans, electronic equipment, and military targets generally, and the intensity depends on the specific technology used in an arsenal and also the purpose.

The heart of Allan Memorial's Radio Telemetry Laboratory (a telemeter is an electrical apparatus for measuring a quantity, transmitting the result by radio to a distant station, and there indicating or recording it) was called the Grid Room. In the Grid Room, an involuntary subject would be strapped into a chair, by force if necessary. Violent resistance was quelled with curare, the powerful plant extract used in arrow poisons by South American Indians and in medicine to produce muscular paralysis. From a head bristling with electrodes and transducers, the subdued subject's brain waves would be beamed to a nearby reception room crammed with voice analyzers and radio receivers cobbled together by laboratory assistant Leonard Rubenstein. Rubenstein, a man who lacked professional medical credentials, believed passionately in the political uses of mind control. In experiments at Allan Memorial's telemetry lab, he declared, would "one day help governments will keep tabs on people without their knowledge.

CHAPTER EIGHT

"To the past or to the future. To an age when thought is free. From the Age of Big Brother, from the Age of the Thought Police, from a dead man... greetings."

- George Orwell (1984)

Scientific research continued within the United States and globally. As early as the 1920s, European scientists made discoveries which paved the way for future development of brain stimulation and brain stimulating technology. The Swiss W. R. Hess could identify 4,000 different places in the brain's hypothalamus, which are in direct contact to certain physical and mental reactions. By stimulating specific points in the brain by an electrical current, the stimulation of one point of the brain could bring about aggressive reactions, while the stimulation of another point could bring about calmness. Through electrical currents to the brain, Dr. Hess could change peoples' personalities, bring about feelings of happiness or sadness, hunger or satisfaction, etc. All of this was achieved over seventy years ago.

Working in Germany during the 1920s, Hans Berger, a psychiatrist, developed the Human Electroencephalograph (EEG – brainwaves). It's important application from the 1930s onwards was initially in the field of Epilepsy. The EEG revealed the presence of electrical discharges in the brain. It also showed different patterns of brainwave discharges associated with different seizure types. The EEG also helped to locate the site of seizure discharges and expanded the possibilities of neurosurgical treatments, which became much more widely available from the 1950s onwards in London, Montreal and Paris. The development of the electroencephalograph (EEG)—an apparatus for detecting and recording brain waves—offered brain physiologists the key to unlock the mysteries of the body's pivotal organ of thought, intellect and personality and set the stage for psychological electronic technology, which would enable man to eventually read human thought combining electromagnetic energy and microwave radio wave with various other technical application.

The Soviets reportedly began to delve into the actual biological effect of microwaves as early as 1953. A number of laboratories were set up across the Soviet Union and in the Eastern Europe, including one at the Institute of Hygiene and Occupational Diseases Academy of Medical Sciences. Although the Soviets reported on their experiments in the open literature, the parameters they defined were insufficient for duplicating the experiments, and some scientists in the United States questioned whether the whole matter was disinformation. It was not. The Russian experiments in the control of a person's mind, factually, through hypnosis and radio waves were conducted in the 1930s.

The history of science, the military, corporations and government involvement from the 1940s up to today says that classified military research was used to fight The Cold War. For this reason, it can be seen how secret, highly classified technology could have been developed outside the public eye and with the "Top Secret" support of top government officials. Teams of elite scientists were used to tackle military problems resulting from World War II (1939 to 1945.) The Cybernetics Group, Brain Research Institute, the Institute for Defense Analysis and their JASON Group and also their Golden Fleece Group are a few examples. Prominent scientists were given military funding generated by the race to surpass the Russians during The Cold War (1946 to 1991) immediately following World War II.

Today, it is commonly known in the scientific community of electromagnetic warfare, and microwave technology that the United States government woke to the reality of psychotropic (psychological electronic) when from 1960 to 1965 the American Embassy in Moscow was targeted by a mixture of electromagnetic and microwaves which caused a wide range of physical and technologically induced mental illnesses among United States personnel serving there. In 1965, the Department of Defense (DOD) discovered that the American embassy in Moscow was being purposely irradiated by the Russians with massive levels of microwaves.

Today, microwaves technology is the most dangerous form of electronic harassment and is quite easy to implement against a person. Directed harmful high energy microwave devices and weapons can include such devices and equipment that are easily available almost anywhere in the World where there are people. Microwave ovens can be modified to focus and direct up to 1200 watts or more, depending on the model, of microwave energy at a person or property and are easy to produce. The results of an attack by a high energy device can be lethal. They can kill or at the very least disable a human being or other animals.

By that time, the DOD's secretive Advance Research Projects Agency (ARPA) now DARPA (Defense) at the Walter Reed Army Institute of Research in the nation's capital had itself developed a prodigious arsenal of electromagnetic weapons. Doctor José Delgado—whose current work with radio waves was underwritten by the CIA and Navy—believed scientists could transform, shape, direct and robotize humankind. "The great danger of the future," Delgado warned, "is that we will have robotized human beings who are not aware that they have been robotized."

Dr. José Manuel Rodriguez Delgado (born August 8, 1915) is a Spanish professor of physiology at Yale University, famed for his research into mind control through electrical stimulation of regions in the brain. He also stated, in a speech recorded in the February 24, 1974 edition of the CONGRESSIONAL RECORD No. 26, Vol. 118, Dr. Delgado had this to say:

"We need a program of psychosurgery for political control of our society. The purpose is physical control of the mind. Everyone who deviates from the given norm can be surgically mutilated. "The individual may think that the most important reality is his own existence, but this is only his personal point of view. This lacks historical perspective. "Man does not have the right to develop his own mind. This kind of liberal orientation has great appeal. We must electrically control the brain. Someday, armies and generals will be controlled by electric stimulation of the brain".

The microwave rays directed at the Embassy and eventual death of the U.S. Ambassador from adult leukemia, and two others from cancer was surely a wakeup call to microwave as a very important weapon. The Embassy personnel's deaths were later traced to being a direct result of Soviet directed microwave technology radiation from a nearby building from the Embassy. These high-powered beams also resulted in the approximate 1800 U.S. State Department Embassy personnel working there later being compensated for various health issues by the United States Government.

There were an unusually large and coincidental number of illnesses reported among the residents of the compound. U.S. Ambassador Walter Stoessel developed a rare blood disease similar to leukemia; he was suffering headaches and bleeding from the eyes. A source at the State Department informally admitted that excessive radiation had been leaking from his telephone; an American high frequency radio transmitter on the roof of the building had, when operating, induced high frequency signals well above the U.S. safety standard through the phones in the political section, as well as in lines to Stoessel's office. No doubt, National Security Agency or CIA electronic devices also contributed to the electromagnetic environment at the embassy, although values for these were never released, as they are secret. Stoessel was reported as telling his staff that the microwaves could cause leukemia, skin cancer, cataracts and various forms of emotional illness. White blood cell counts were estimated to be as high as 40% above normal in one third of the staff, and serious chromosome damage was uncovered...

The Soviets began research on biological effects of microwaves in 1953. A special laboratory was set up at the Institute of Hygiene and Occupational Diseases, Academy of Medical Sciences. Other labs were set up in the U.S.S.R. and in Eastern Europe that study both effects of microwaves and low frequency electromagnetic radiation.

Dr. Stefan Thomas Poisson, an Austrian born U.S. economist and military strategist, (March 15, 1913 in Vienna - April 26, 1995) one of "the greatest strategic philosophers of the 20th Century", founder of International Strategic Studies Association, former member of Mankind Research Unlimited and former psychological warfare expert with the Office of Naval Research in an article entitled, "Scientific Advances Hold Dramatic Prospects for Psy-Strategy" by Possony, Stefan (1983, July), Defense & Foreign Affairs, P.34., Dr. Possony stated that mind control is feasible and militarily important. The article is an overview of the Soviet-U.S. race to develop mind control technology and a precise explanation for why there is illegal experimentation going on to this day. Dr. Possony further states, "The target cannot be persuaded to listen. It is the other way around; he may listen if he already is fully or partially persuaded, and if the program is attractive in addition to informative, and if it helps him in his activities." Dr. Possony was also a one-time Science Advisor to the Department of Defense, who conceived the U.S. Strategic Defense Initiative, and surrounding the U.S. Embassy in Moscow is reported as saying:

"After the death of our ambassador in Moscow, due to contracting leukemia, and a couple of other employees, it suddenly dawned on us to have a real careful look at what was happening there." As a result, a huge project got underway in the United States to catch up with the Soviet advantage of many years before them specifically in the area of remote sensing and stimulation."

The Defense Intelligence Agency (DIA) is a member of the Intelligence Community of the United States, and is a major producer and manager of military intelligence for the United States Department of Defense, employing over 16,500 military and civilian employees worldwide. The DIA, designated in 1986 as a Defense Department combat support agency, was established in 1961 as a result of a decision by Secretary of Defense Robert S. McNamara, under President John F. Kennedy. DIA is a member of the United States

Intelligence Community, reporting to the Director of National Intelligence. The Director of National Intelligence (DNI), is the United States government official subject to the authority, direction and control of the President, who is responsible under the Intelligence Reform and Terrorism Prevention Act of 2004 for:

Serving as the principal advisor to the President, the National Security Council, and the Homeland Security Council for intelligence matters related to national security;

Serving as the head of the sixteen-member Intelligence Community; and Overseeing and directing the National Intelligence Program.

The Intelligence Reform and Terrorism Prevention Act of 2004 created the office of the Director of National Intelligence (DNI), which took over management and leadership of the Intelligence Community. Prior to establishment of the DNI, the head of the Intelligence Community was the Director of Central Intelligence (DCI). The DCI concurrently served as the Director of the Central Intelligence Agency (CIA). The Central Intelligence Agency (CIA) is a civilian intelligence agency of the United States government, reporting to the Director of National Intelligence, responsible for providing national security intelligence assessment to senior United States policymakers. The CIA also engages in covert activities at the request of the President of the United States.

A Defense Intelligence Agency review of Communist literature affirmed microwave sound and indicated voice transmission. The report states, "Sounds and possibly even words which appear to be originating intra-cranially (within the head) can be induced by signal modulation at very low average power densities." Among microwave weapon implications are "great potential for development into a system for disorientating or disrupting the behavior patterns of military or diplomatic personnel." An Army Mobility Equipment Research and Development Command report affirms microwave speech transmission with applications of "camouflage, decoy, and deception operations." "One decoy and deception concept presently being considered is to remotely create noise in the heads of personnel by exposing them to low power, pulsed microwaves. By proper choice of pulse characteristics, intelligible speech may be created" quotes the report.

"The CIA inquired whether I thought electromagnetic radiation beamed at the brain from a distance could affect the way a person might act, and if microwaves could be used to facilitate brainwashing or to break down prisoners under investigation." The State Department elected to keep the so-called Moscow Signal a secret from American Embassy employees—and studied the side-effects of the radiation instead. Ambassador Walter J. Stoessel Jr., a long-time American diplomat in the Soviet Union, whose office was situated in the magnetic beam's center, succumbed by stages to blood disease, bleeding eyes, nausea and lymphoma. State Department employees Charles Bohlen and Llewellyn Thompson fell prey to cancer. The existence of the Soviet beam was finally acknowledged by the U.S. in 1976, in response to a report by syndicated columnist Jack Anderson. Officially, the State Department concluded that the microwave saturation of the embassy served not to brainwash, but to activate bugging devices in the walls. However, Dr. Zaret, after conducting his own tests, deduced that the Moscow Signal was psychoactive. "Whatever other reasons the Russians may have had for irradiating the American embassy," posits Zaret, "they believed the beam would modify the behavior of personnel."

Discovered in 1962, these complex set of electromagnetic signals, 10 years later were investigated by the CIA, which hired a consultant, Milton Zaret, and code named the research "Project Pandora".

The Project PANDORA, included a number of parallel projects, such as Project's TUMS, MUTS, and BAZAR, involving the CIA, Advanced Research Project Agency (ARPA) again, today, now called DARPA. DARPA was created as the Advanced Research Projects Agency, by Public Law 85-325 and Department of Defense Directive 5105.15, in February 1958. Its creation was directly attributed to the launching of Sputnik and to U.S. realization that the Soviet Union had developed the capacity to rapidly exploit military technology. Additionally, the political and defense communities recognized the need for a high-level Department of Defense organization to formulate and execute Research & Development (R&D) projects that would expand the frontiers of technology beyond the immediate and specific requirements of the military services and their laboratories. In pursuit of this mission, DARPA has developed and transferred technology programs encompassing a wide range of scientific disciplines which address the full spectrum of national security needs.

Other parallel projects were run by the State Department, the Navy and the Army. They were tasked to study the effects of the emitted Soviet microwaves on animals and humans. The electromagnetic "Moscow Signals," which had each day targeted the U.S. Embassy in Moscow. These signals in the short 'S' and long 'L' spectrum had complex modulations with a pattern of variations, some of which were random. A Top Secret-Eyes Only memorandum, dated 20 December 1966 from ARPA, shows the significance of this project. (Various excerpt from: The Military Use of Electromagnetic Microwave and Mind Control Technology by Dr. Armen Victorian, extract from: Lobster Magazine.)

In the mid-1970s Dr. Joseph C. Sharp helped to develop microwave hearing technology for DARPA conducting research at Walter Reed Army Hospital. Dr. Sharp, was also a *Project Pandora researcher at Walter Reed Army Institute of Research and was involved in work in areas that were so secret that he couldn't even tell his boss. He is documented as 'one' of the first to conduct experiments in which the human brain received a message carried to it by microwave transmission in the United States. Dr. Sharp was also able to recognize spoken words that were modulated on a microwave carrier frequency by an "audiogram", an analog of the words' sound vibrations, and carried into his head inside a chamber where he sat.

There were three scientists who pioneered the work of using an electromagnetic field to control human behavior. These three were Dr. Jose Delgado, psychology professor at Yale University; Dr. W. Ross Adey, a physiologist at the Brain Research Institute at UCLA; and Dr. Wilder Penfield, a Canadian. Dr. Adey's research at the Brain Research Institute of the University of California was funded by the Central Intelligence Division. In their Project Pandora a catalogue of different brain signals for specific actions, emotions and pathological states of mind were recorded. The research further verified that when microwaves were used to fire these signals at victims' brains, they experienced the moods, behavior, and the pathological states, carried by the signals. This meant that by mimicking natural brain frequencies, the human brain could be controlled remotely by use of extremely low frequency broadcast carried by pulse modulated microwave beams (ELF pulse modulated microwave remote mind control technology,) Microwave and Control by Tim Rifat.

Not only were studies taking place in the United States including institutions of higher learning, but similar studies were also being conducted in Great Britain during this time. In the Washington Associated Press, May 22, 1988, Barton Reppert, Associated Press, Writer, "Looking at the Moscow Signal the Zapping of an Embassy 35 years later The Mystery Lingers." Reppert stated "Since the early 1980s, however, federal government support for non-ionizing radiation bio effects has declined markedly. However, W. Ross Adey, a leading researcher based at the Veterans Administration Medical Center in Loma Linda, California, later told a House Subcommittee that current levels of government funding now about $7 million a year are disastrously low." In the 1980s Dr. Adey performed some crucial experiments using microwave carrier-waves modulated with ELF waves to modify brain tissue responses and global research in these areas beginning all over the world.

Logically, in order for any mind control technology to be effective and successful, the operators, it would seem would need to be neutralized and controlled first. The TETRA System is designed to ensure that law enforcement is in sync with mind control efforts in case of a global uprising. In order for this to happen, they must first be successfully mind controlled themselves. Based on my experience with law enforcement in Los Angeles, California beginning in 2005, I had to consider this as a possible reason that the group Targeting me, refused to accept, two things, one, that I was not scared, and two, in lieu of their inability to substantiate any wrongful allegations against me, to move on as they lingered around the clock with continued senseless, merciless abuse.

In 2001 – The TETRA System: Mass UK Mind Control Technology and the Zombification of Britain's Police is now a Reality by Tim Rifat, excerpts below reports:

"The TETRA system pulses at 17.6 Hz broadcast at 400 MHz are essentially the Pandora Project funded by the CIA in the late 1960s and early 1970s. Dr. Ross Adey, the chief researcher on the Pandora Project released a video to leading United Kingdom researchers which proved that not only does the TETRA system cause ELF zombification by massive release of calcium ions in the cerebral cortex and the nervous system, but the activated calcium ions also cause massive hormonal disturbances which lead to frenzied imbalances, emotional and physical states.

Use of the TETRA system by the police will lead to psychotronically controlled police officers who may be totally controlled in any situation and are very useful for states of economic or social chaos where extreme and violent behavior is needed without any conscious or moral compunction - so-called police robots.

So, sophisticated is this research, and I refer to Operation Pandora Joint CIA/MI6 Operation since the 1960s, Operation Woodpecker USSR 1976 (Russian version of HAARP), Operation HAARP still running in USA work along with GWEN Towers; they are able to define specific pulse frequencies to cause specific brain malfunctions or illnesses. For instance: cause illness, depression/suicide, manic behaviour, and anger, blindness if aimed at the head or heart attack if aimed at the chest. Other consequences of frequencies used but not listed here are hysteria, trauma, lust, murder and cancer, and may all be induced. (Confidential Report on TETRA for the Police of England and Wales by Barrie. Trower.) More information can be found in "Update on Murder and Mind Control: The Secret Uses of ELF Modulated Microwave and radio frequency (RF) by the British Army" regarding the Great Britain.

The CIA-funded Dr. Ross Adey to investigate the mind-controlling and hormonal-effecting uses of pulse and amplitude-modulated microwave and RF. Ross Adey experimented with 450MHz, which in the UK system is microwave, as the British take 400MHz to 400GHz as microwave. His associate, Dr. Blackmore, experimented with RF frequencies at around 150MHz. This was developed by the British Army, the Secret Police, into a variety of pulse-modulated, or amplitude modulated, radio frequency or microwave transmitters, which focused on the target ELF which had bioactive effects to kill or mind control their targets. The amplitude or pulse modulation of the carrier wave allows the British Army/Secret Police (MI5) operatives to induce ELF frequencies on the victim, even though the carrier wave is in the RF or microwave range."

Of course, typical of anyone speaking up, on these subjects, Mr. Rifat has also been labelled delusional and self-serving by those seeking to discredit his exposure efforts. Before reading this, I had no idea how accurate I had been when I said jokingly said that those operating around me had been effectively mind controlled themselves. As a result, when I stumbled on the TETRA system research, I was not surprised due to their behaviour and inability to cease their efforts in light no success or perception of immorality or wrongdoing. They had crossed this fine line a ways back.

Today, laws such as USCA, Title 18, Chapter 119 or Executive Order 12333 now would allow government, through the authorization of a cognizant court, to utilize multiple avenues towards the assurance of an effective demonstration of these technologies now delivered by satellite radar or laser in the United States and globally? Who would have guessed that as a result of the deployment of commercial communications satellites around the globe beginning in the late 50s and heavily in the 1970's, that microwave technology could emanate from space for testing would be, be conducted by a government legally against its citizens or the unsuspecting.

CHAPTER NINE

"Reality exists in the human mind, and nowhere else. Not in the individual mind which can make mistakes, and in any case soon perishes: only in the mind of the Party, which is collective and immortal."

- George Orwell (1984)

TIMELINE TWO

Archived as:

http://www.stealthskater.com/Documents/iopaea_1.doc [pdf]

More at physics: http://www.stealthskater.com/Science.htm

NOTE: because important websites are frequently "here today but gone tomorrow" the following was archived from:

http://www.cassiopaea.org/cass/timeline.htm on November 3, 2002.

This is NOT an attempt to divert readers from the aforementioned website. Indeed, the reader should only read this back-up copy if it cannot be found at the original author's site.

This timeline -- prepared by a researcher ["JH"] of our Quantum Future School with many linked sources -- barely scratches the surface. It is our hope that readers will do additional research and provide us with more links and connections to this spider web of

Cosmic COINTELPRO that has blanketed the Earth with lies, deception, confusion, and tricks and traps - the magnets of impending Global Destruction. See also Star of Sorcerer's for additional connections.

A far more extensive timeline could be created by including the information from Freddy Silva's book on Crop Circles, Richard Dolan's book on UFOs and the National Security State, and the research included in the Adventures Series. The reader will also want to read "To Be or Not to Be" for more background. We will continue to work on the project in hopes that by seeing the various threads together, more people will realize just how it all connects and how totally we have been duped, and how evil the plans of the Controllers truly are.

Laura Knight-Jadczyk & Dr. Arkadiusz Jadczyk
http://www.cassisopaea.org

SECRET GOVERNMENT, LSD, Esalen, HAARP, and the Cosmic COINTELPRO or "When You Dance with the Devil..."

1931

Dr. Cornelius Rhoads - under the auspices of the Rockefeller Institute for Medical Investigations - infects human subjects with cancer cells. He later goes on to establish the U.S. Army Biological Warfare facilities in Maryland, Utah, and Panama, and is named to the U.S. Atomic Energy Commission. While there, he begins a series of radiation exposure experiments on American soldiers and civilian hospital patients.

1932

The Tuskegee Syphilis Study begins. 200 black men diagnosed with syphilis are never told of their illness, are denied treatment, and instead are used as human guinea pigs in order to follow the progression and symptoms of the disease. They all subsequently die from syphilis. Their families were never told that they could have been treated.

1933

A Humanist Manifesto is published with 34 prominent signatories at the time.

1934

A. "A method for Remote Control of Electrical Stimulation of the Nervous System", a monograph by Drs. E.L. Chaffee and R.U. Light.

B. Experiments in Distant Influence, a book by Soviet Professor Leonid L. Vasiliev. He also wrote the article "Critical Evaluation of the Hypnogenic Method" concerning the work of Dr. I. F. Tomashevsky on experiments in remote control of the brain.

1935

The Pellagra Incident

http://www.mail-archive.com/ugandanet@kym.net/msg10247.html

After millions of individuals die from Pellagra over a span of 2 decades, the U.S. Public Health Service finally acts to stem the disease. The director of the agency admits it had known for at least 20 years that Pellagra is caused by a niacin deficiency, but failed to act since most of the deaths occurred within poverty-stricken black populations.

1940

400 prisoners in Chicago are infected with malaria in order to study the effects of new and experimental drugs to combat the disease. Nazi doctors later on trial at Nuremberg cite this American study to defend their own actions during the Holocaust.

1942

Chemical Warfare Services begins mustard gas experiments on approximately 4,000 servicemen. The experiments continue until 1945 and made use of Seventh Day Adventists who chose to become human guinea pigs rather than serve on active duty.

1943

In response to Japan's full-scale germ warfare program, the U.S. begins research on biological weapons at Fort Detrick, MD.

1944

U.S. Navy uses human subjects to test gas masks and clothing. Individuals were locked in a gas chamber and exposed to mustard gas and lewisite.

1945

- A. After World War II, the Allies discovered the Japanese had been developing a "death ray" utilizing very short radio waves focused into a high-power beam. Tests were done on animals. The Japanese denied ever testing it on humans. (From the Strategic Bombing Survey, Imperial War Museum, London. Cited with photocopies in "Japanese Death Ray", by Peter Lewis, Resonance#11, pp 5-9)

- B. Project Paperclip is initiated. The U.S. State Department, Army intelligence, and the CIA recruit Nazi scientists and offer them immunity and secret identities in exchange for work on top-secret government projects in the United States.

C. "Program F" is implemented by the U.S. Atomic Energy Commission (AEC). This is the most extensive U.S. study of the health effects of fluoride, which was the key chemical component in atomic bomb production. One of the most toxic chemicals known to man, fluoride causes marked adverse effects to the central nervous system. But much of the information is squelched in the name of "national security" because of fear that lawsuits would undermine full-scale production of atomic bombs.

1946

Patients in VA hospitals are used as guinea pigs for medical experiments. In order to allay suspicions, the order is given to change the word "experiments" to "investigations" or "observations" whenever reporting a medical study performed in one of the Nation's veteran's hospitals.

1947

A. Colonel E.E. Kirkpatrick of the U.S. Atomic Energy Commission issues a secret document (Document 07075001, January 8, 1947) stating that the agency will begin administering intravenous doses of radioactive substances to human subjects.

B. The CIA begins its study of LSD as a potential weapon for use by American intelligence. Human subjects (both civilian and military) are used with-and-without their knowledge.

1950

A. The Department of Defense begins plans to detonate nuclear weapons in desert areas and monitor downwind residents for medical problems and mortality rates.

B. In an experiment to determine how susceptible an American city would be to biological attack, the U.S. Navy sprays a cloud of bacteria from ships over San Francisco. Monitoring devices are situated throughout the city in order to test the extent of infection. Many residents become ill with pneumonia-like symptoms.

C. The French conducted research on infrasonic weapons (from "The Road from Armageddon", by Peter Lewis, Resonance#13, pp 9-14).

D. The newly-formed CIA initiated studies in mind-control programs in 1950 with Project Bluebird (rechristened "Artichoke") in 1951. To establish a ' cover story' for this research, the CIA funded a propaganda effort designed to convince the World that the Communist Bloc had devised insidious new methods of re-shaping the human will. The CIA's own efforts could therefore - if exposed - be explained as an attempt to "catch up" with Soviet and Chinese work.

The primary promoter of this 'line' was one Edward Hunter, a CIA contract employee operating undercover as a journalist and - later - a prominent member of the John Birch society.

Hunter offered 'brainwashing' as the explanation for the numerous confessions signed by American prisoners of war during the Korean War and (generally) UN-recanted upon the prisoners' repatriation. These confessions alleged that the United States used germ warfare in the Korean conflict - a claim which the American public of the time found impossible to accept.

Many years later, however, investigative reporters discovered that Japan's germ warfare specialists (who had wreaked incalculable terror on the conquered Chinese during WWII) had been mustered into the American national security apparatus. And the knowledge gleaned from Japan's horrifying germ warfare experiments probably WAS used in Korea just as the 'brainwashed' soldiers had indicated. Thus, we now know that the entire brainwashing scare of the 1950s constituted a CIA hoax perpetrated upon the American public.

CIA deputy director Richard Helms admitted as much when in 1963, he told the Warren Commission that "Soviet mind-control research consistently lagged years behind American efforts."

1951

A. Alfred Hubbard first tries LSD. An OSS officer in WWII, Hubbard first took LSD in 1951 and proceeded to "turn on" several individuals prominent in LSD research including Dr. Humphrey Osmond, Myron Stolaroff, and Aldous Huxley, earning him the title of "the Johnny Appleseed of LSD" (Lee, Martin and Schlain, Bruce, Acid Dreams, Grove Press, 1985, pg. 44). Circa 1951, Hubbard later did undercover work for several agencies including the FDA and FBI. He reportedly tried (and failed) to "turn on" J. Edgar Hoover. He introduced LSD to many high-ranking intelligence officers. In the early 1950s, he refused an offer to join the CIA (Lee and Schlain, pg. 52). In all, it is estimated that Hubbard introduced LSD to over 6,000 individuals. He worked until 1965 at the International Foundation for Advanced Study (mis-identified here, I think, as the International Federation for Advanced Studies) (Fahey, Todd Brendan, The Original Captain Trips", High Times, November 1991). Fahey describes Hubbard's work at SRI differently, placing him with the Alternative Futures Project which sought to "turn on" the World's political and business leaders. He left SRI in 1974 and died on August 31, 1982 (Fahey).

B. The Department of Defense begins open air tests using disease-producing bacteria and viruses. Tests last through 1969 and there is concern that people in the surrounding areas have been exposed.

1952

C. As a child in 1952, Jack Sarfatti claims to have received phone calls from the mechanical voice of a conscious computer aboard a spaceship, recruiting him along with 400 others for some special project. These calls have similarities to the mechanical voice which talked to Andrijah Puharich via his tape recorder. Sarfatti was later associated with Puharich. Puharich first contacts "The Nine" - a group of channeled being via a medium.

D. During the CIA's MK-ULTRA mind-control program, John Lilly briefed the intelligence community on his work to map out the brains of animals using implanted electrodes. He abandoned this line of work because he felt it was unethical. John Lilly studied the effects of sensory deprivation tanks and also briefed the intelligence community with his progress. Lilly refused to let any of his work be classified and ended up leaving the National Institute of Health when he found that he could not work without the interference of the Government.

E. Project Moonstruck/CIA:

Electronic implants in brain and teeth

Targeting: Long range

Implanted during surgery or surreptitiously during abduction

Frequency range: HF - ELF transceiver implants

Purpose: Tracking, mind & behavior control, conditioning, programming, covert operations

Functional Basis: Electronic Stimulation of the Brain (E.S.B.)

1953

A. John C. Lilly - when asked by the director of the National Institute of Mental Health (NIMH) to brief the Central Intelligence Agency (CIA), Federal Bureau of Investigation (FBI), National Security Agency (NSA), and the various military intelligence services on his work using electrodes to stimulate directly the pleasure and pain centers in the brain - refused. He said,

"Dr. Antione Redmond - using our techniques in Paris - has demonstrated that this method of stimulation on the brain can be applied to the human without help of the neurosurgeon ... This means that anybody with the proper apparatus can carry this out covertly with no external signs that electrodes have been used in that person. I feel that if this technique got into the hands of a secret agency, they would have total control over a human being and be able to change his beliefs extremely quickly, leaving little evidence of what they had done."

(from "Mind Control and the American Government", by Martin Cannon in Lobster#23, pp 2-10. Cannon quotes Lilly from his book The Scientist, Berkeley, Ronin publishers, 1988, also Bantam Books 1981. Research by Peter Lewis.)

[note: After a statement like that of Dr. Lilly's, how long do you think it would take the agencies, FBI, CIA, NSA, etc. to contact Dr. Redmond in Paris?]

B. Project MK-ULTRA/ CIA:

Drugs, electronics and electroshock

Targeting: Short range

Frequencies: VHF HF UHF modulated at ELF

Transmission and Reception: Local production

Purpose: Programming behavior, creation of "cyborg" mentalities

Effects: narcoleptic trance, programming by suggestion

Subprojects: Many.

Pseudonym: Project Artichoke

Functional Basis: Electronic Dissolution of Memory, E.D.O.M. (Disinfo ???)

When the CIA's mind-control program was transferred from the Office of Security to the Technical Services Staff (TSS) in 1953, the name changed again to MK-ULTRA.

Later still, in 1962, mind-control research was transferred to the Office of Research and Development; project cryptonyms remain unrevealed. What was studied? Everything including hypnosis, conditioning, sensory deprivation, drugs, religious cults, microwaves, psychosurgery, brain implants, and even ESP. When MK-ULTRA "leaked" to the public during the great CIA investigations of the 1970s, public attention focused most heavily on drug experimentation and the work with ESP.

Mystery still shrouds another area of study - the area which seems to have most interested ORD: psychoelectronics

C. Martin Cannon, The Controllers: A New Hypothesis of Alien Abduction:

"The MK-ULTRA program was a covert behavior modification program run by the CIA in the early 1950s with the purpose of finding ways to make men more suggestible and involving the use of pain, drugs, and hypnosis on unsuspecting human guinea pigs.

D. The first person to publicly expose the CIA's use of "pain-drug-hypnosis" was L. Ron Hubbard, the founder of Scientology who wrote in his 1951 book Science of Survival that it had become so extensively employed in espionage work that it was long past the time that people should have become alarmed about it.

"Mr. Hubbard's statement was found to be true in the 1970s when the CIA's program became public knowledge after the Freedom of Information Act enabled investigators to document the agency's inhumane and grotesque experiments on human subjects.

The ensuing outcry over the use of mind-bending drugs - which combined with electric shock caused the deaths or maiming of untold numbers of people - drew comparisons between the CIA and the infamous Nazi doctors and led to Congressional hearings into the intelligence agency." - an40286@anon.penet.fi (probably from the Scientology Guardians Organization).

E. U.S. military releases clouds of zinc cadmium sulfide gas over Winnipeg, St. Louis, Minneapolis, Fort Wayne, the Monocacy River Valley in Maryland, and Leesburg, Virginia. Their intent is to determine how efficiently they could disperse chemical agents.

F. Joint Army-Navy-CIA experiments are conducted in which tens-of-thousands of people in New York and San Francisco are exposed to the airborne germs Serratia marcescens and Bacillus glogigii.

1955

A. (circa) Dr. Louis West, friends with Aldous Huxley. It was Huxley who suggested that West combine LSD and hypnosis in his experiments (Lee, Martin, and Schlain, Bruce, Acid Dreams, Grove Press, 1985, pg. 48). West was an Air Force Major, chairman of the Psychiatry Department of UCLA, director of the Neuro-Psychiatric Institute, and an expert in hypnosis. West was a veteran of the CIA's MK-ULTRA mind-control program and worked on interrogation techniques using hypnosis and LSD. West once killed an elephant by grossly overestimating a dose of LSD (elsewhere, I have heard that the tranquilizers required to calm the animal caused its death). West also studied the returning American POWs from Korea for the effects of brainwashing (Scheflin, Alan and Opton, Edward Jr., The Mind Manipulators, Paddington Press Ltd, 1978, pg. 149-50).

B. Morris K. Jessup published The Case For the UFO.

C. The CIA - in an experiment to test its ability to infect human populations with biological agents - releases a bacteria withdrawn from the Army's biological warfare arsenal over Tampa Bay, Fl.

D. Army Chemical Corps continues LSD research, studying its potential use as a chemical incapacitating agent. More than 1,000 Americans participate in the tests, which continued until 1958.

1956

U.S. military releases mosquitoes infected with Yellow Fever over Savannah, GA and Avon Park, FL. Following each test, Army agents posing as public health officials test victims for effects.

1957

It has now been documented that millions of doses of LSD were produced and disseminated under the aegis of the CIA's Operation MK-ULTRA. LSD became the drug of choice within the agency itself, and was passed out freely to friends of the family including a substantial number of OSS veterans.

For instance, it was OSS Research and Analysis Branch veteran Gregory Bateson who "turned on" the Beat poet Allen Ginsberg to a U.S. Navy LSD experiment in Palo Alto, California. Not only Ginsberg but also novelist Ken Kesey and the original members of the Grateful Dead rock group opened the doors of perception courtesy of the Navy. The guru of the 'psychedelic revolution' - Timothy Leary - first heard about hallucinogens in 1957 from Life magazine (whose publisher Henry Luce was often given Government acid like many other opinion shapers), and began his career as a CIA contract employee.

At a 1977 "reunion" of acid pioneers, Leary openly admitted, "everything I am, I owe to the foresight of the CIA." [Michael J. Minnicino, "The New Dark Age, The Frankfurt School, and 'Political Correctness'", Fidelio, v1 #1] 1958

A. Project Argus

Between August and September 1958, the US Navy exploded 3 fission-type nuclear bombs 480 km above the South Atlantic Ocean in the part of the lower Van Allen Belt closest to the Earth's surface. In addition, 2 hydrogen bombs were detonated 160 km over Johnston Island in the Pacific.

The military called this "the biggest scientific experiment ever undertaken". It was designed by the U.S. Department of Defense and the U.S. Atomic Energy Commission, under the code name 'Project Argus'. The purpose appears to be to assess the impact of high-altitude nuclear explosions on radio transmission and radar operations because of the electromagnetic pulse (EMP), and to increase understanding of the geomagnetic field and the behavior of the charged particles in it. This gigantic experiment created new (inner) magnetic radiation belts encompassing almost the whole Earth and injected sufficient electrons and other energetic particles into the ionosphere to cause worldwide effects. The electrons traveled bac- and-forth along magnetic force lines, causing an artificial "aurora" when striking the atmosphere near the North Pole.

B. The U.S. Military planned to create a "telecommunications shield" in the ionosphere, reported in 13-20 August 1961, Keesings Historisch Archief (K.H.A.). This shield would be created "in the ionosphere at 3,000 km height by bringing into orbit 350,000 million copper needles, each 2-4 cm long [total weight 16 kg], forming a belt 10 km thick and 40 km wide, the needles spaced about 100 m apart." This was designed to replace the ionosphere "because telecommunications are impaired by magnetic storms and solar flares."

The U.S. planned to add to the number of copper needles if the experiment proved to be successful. This plan was strongly opposed by the Intentional Union of Astronomers.

C. Project Orion/ USAF:

Drugs, hypnosis, and ESB

Targeting: Short range, in person

Frequencies: ELF Modulation

Transmission and Reception: Radar, microwaves, modulated at ELF frequencies

Purpose: Top-security personnel debriefing, programming, insure security and loyalty

Pseudonym: "Dreamland"

[Stealth Skater note: Bob Lazar said that he was ordered to take drugs that smelled like "pine" as part of his clearance to the S4 projects.]

D. While Lilly implies that he left the NIH because of unethical government interference, his Communications Research Institute (founded in the 1958 to study dolphins) was partially funded by the Air Force, NASA, NIHM, the National Science Foundation, and the Navy. He was assisted in this work by Gregory Bateson.

While experimenting with sensory deprivation and LSD and ketamine, Lilly came to believe that he was in psychic contact with the aliens of what he called the "Earth Coincidence Control Office". The aliens were guiding events in Lilly's life to lead him to work with dolphins, which were psychic conduits between aliens and humans. The aliens are acting for the survival of organic lifeforms against artificial intelligences called "solid-state lifeforms". (E) LSD is tested on 95 volunteers at the Army's Chemical Warfare Laboratories for its effect on intelligence.

1959

Huxley speeches in London on "Latent Human Potential". COINTELPRO is kicked off and the games begin. 1960

MK-DELTA. CIA:

> Fine-tuned electromagnetic subliminal programming
>
> Targeting: Long Range
>
> Frequencies: VHF HF UHF Modulated at ELF
>
> Transmission and Reception: Television antennae, radio antennae, power lines, mattress spring coils, modulation on 60-Hz wiring.
>
> Purpose: programming behavior and attitudes in general population
>
> Effects: fatigue, mood swings, behavior dysfunction, and social criminality
>
> Pseudonym: "Deep Sleep", R.H.I.C.

A. Hal Puthoff - according to author Jim Schnabel (and confirmed by Dr. Puthoff) - served at the NSA in the early 1960s during his tour with the Navy (not the Army as McRae reported) and later stayed on as a civilian. Joined SRI in 1971 as a specialist in laser physics. Served as an officer in the Navy from 1960-63 at Ft. Meade.

B. Headlines read,

> "Khrushchev Says Soviets Will Cut Forces a Third; Sees 'Fantastic Weapon' ". (From article of same title, by Max Frankel, New York Times, Jan. 15, 1960, p.1 as cited in "Tesla's Electromagnetics and Its Soviet Weaponization", paper by T.E. Bearden.)

C. The International Foundation for Advanced Study (IFAS) is established. Founded by Myron Stolaroff and Paul Kurtz and located in Menlo Park, California. Studied the effects of LSD and mescaline from 1961 to 1965. (Anderson, Walter Truett, The Upstart Spring, Addison-Wesley Publishing, 1983) The foundation also offered LSD therapy for $500 a session. In late 1961, the foundation released The Psychedelic Experience: A New Concept in Psychotherapy. (Stevens, Jay, Storming Heaven, Atlantic Monthly Press, 1987, pg. 177-9) Also involved with the IFAS were Alfred Hubbard, Vice President Willis Harman, Charles Savage, Robert Mogar, James Fadiman, and Ethel Savage; with Hubbard reportedly supplying the drugs (then legal for research).

D. The Army Assistant Chief-of-Staff for Intelligence (ACSI) authorizes field testing of LSD in Europe and the Far East. Testing of the European population is code named Project "Third Chance"; testing of the Asian population is code named Project "Derby Hat".

1962

A. The Esalen Institute was founded in 1964 by Mike Murphy and Dick Price out of Murphy's family resort. Murphy and Price had been running seminars at the resort beginning in 1962 with speakers gathered through an expanding network of contacts, beginning with Alan Watts, Aldous Huxley, Gregory Bateson, Gerald Heard, and others. [see Anderson, Walter Truett, The Upstart, Addison-Wesley Publishing, 1983 for an expansive history of Esalen] While an engineering professor at Stanford University, Harman led a 1962 conference on human potentiality at the "Esalen Institute of Expanding Vision." Harman went on later to head IONS with Astronaut Edgar Mitchell.

B. Project Starfish

On July 9, 1962, the US began a further series of experiments with the ionosphere. From their description:

"one kiloton device at a height of 60 km, and one megaton and one multi-megaton at several hundred kilometers height" (K.H.A., 29 June 1962). These tests seriously disturbed the lower Van Allen Belt, substantially altering its shape and intensity. "In this experiment, the inner Van Allen Belt will be practically destroyed for a period of time. Particles from the Belt will be transported to the atmosphere. It is anticipated that the Earth's magnetic field will be disturbed over long distances for several hours, preventing radio communication. The explosion in the inner radiation belt will create an artificial dome of polar light that will be visible from Los Angeles" (K.H.A. May 11, 1962). "A Fijian Sailor - present at this nuclear explosion - told me that "the whole sky was on fire" and he thought it would be the End of the World. This was the experiment which called forth the strong protest of the Queen's Astronomer, Sir Martin Ryle in the UK. "The ionosphere [according to the under-standing at that time] - that part of the atmosphere between 65 and 80 km and 280-320 km height - will be disrupted by mechanical forces caused by the pressure wave following the explosion. At the same time, large quantities of ionizing radiation will be released, further ionizing the gaseous components of the atmosphere at this height. This ionization effect is strengthened by the radiation from the fission products... "The lower Van Allen Belt, consisting of charged particles that move along the geomagnetic field lines... will similarly be disrupted. As a result of the explosion, this field will be locally destroyed while countless new electrons will be introduced into the lower belt" (K.H.A. 11 May 1962). "On July 19... NASA announced that as a consequence of the high altitude nuclear test of July 9, a new radiation belt had been formed, stretching from a height of about 400 km to 1600 km; it can be seen as a temporary extension of the lower Van Allen Belt" (K.H.A. August 5,1962).

As explained in the Encyclopedia Britannica:

"... Starfish made a much wider belt [than Project Argus] that extends from low altitude out past L=3 [i.e. three Earth radiuses or about 13,000 km above the surface of the Earth]."

C. Later in 1962, the USSR undertook similar planetary experiments, creating 3 new radiation belts between 7,000 and 13,000 km above the Earth.

According to the Encyclopedia, the electron fluxes in the lower Van Allen Belt have changed markedly since the 1962 high-altitude nuclear explosions by the US and USSR - never returning to their former state.

According to American scientists, it could take many hundreds of years for the Van Allen Belts to destabilize at their normal levels. (Research done by: Nigel Harle, Borderland Archives, Cortenbachstraat 32, 6136 C.H. Sittard, Netherlands.)

1963

Hal Puthoff worked for 8 years in the Microwave Laboratory at Stanford University till 1971

1965

A. "A project in the U.S. called Project Pandora ... was undertaken in which chimpanzees were exposed to microwave radiation.

The man who was in charge of this project said, "the potential for exerting a degree of control on human behavior by low level microwave radiation seems to exist" and he urged that the effects of microwaves be studied for "possible weapons applications". (From "Electromagnetic Pollution: A Little Known Health Hazard. A new means of control?" by Kim Besley, Great Britain, p 14. Research from Woody Blue.) In 1965, Koslov - then a physicist at the Advanced Research Projects Agency (ARPA) - suggested to Charles Weiss (head of security at the State Department) that a "a sober and systematic program of research" look into the "Moscow Signal", which was caused by microwave radiation being beamed into the Moscow American Embassy.

This program eventually evolved into Project Pandora, America's first research program into the possible offensive, anti-personnel use of non-ionizing microwave radiation. (Steneck, Nicholas H., The Microwave Debate, The MIT Press, 1984, pg. 94-5)

B. "Death Ray" weapon was developed by McFarlane Corporation.

Described as a modulated electron gun X-ray nuclear booster, it could be adapted to communications, remote control and guidance systems, EM radiation telemetry, and death ray.

McFarlane claimed NASA stole the patent in 1965. Reported hearings before the House Subcommittee on Department of Defense Appropriations, chaired by Rep. George Mahon (Dem. - Texas). (From "Hearing Voices" by Alex Constantine, Hustler, Jan. 1994, pp. 102-104, 113, 120, 134. Research by Harlan Girard.)

C. The CIA and Department of Defense begin Project MK-SEARCH - a program to develop a capability to manipulate human behavior through the use of mind-altering drugs.

D. Prisoners at the Holmesburg State Prison in Philadelphia are subjected to dioxin - the highly toxic chemical component of Agent Orange used in Viet Nam. The men are later studied for development of cancer which indicates that Agent Orange had been a suspected carcinogen all along.

1966

A. CIA initiates Project MK-OFTEN - a program to test the toxicological effects of certain drugs on humans and animals.

B. U.S. Army dispenses Bacillus subtilis variant niger throughout the New York City subway system. More than a million civilians are exposed when Army scientists drop lightbulbs filled with the bacteria onto ventilation grates.

C. Cleve Backster is a polygraph specialist who helped develop interrogation techniques for the CIA. As of 1986, he ran a polygraph instruction school and the Backster Research Foundation in San Diego. In February 1966, Backster recorded what he believes to be emotional reactions in plants with a polygraph machine. Called the "Backster Effect", the validity of this phenomena is still debated.

1967

CIA and Department of Defense implement Project MK-NAOMI - successor to MK-ULTRA and designed to maintain, stockpile, and test biological and chemical weapons.

1968

A. Eldon Byrd Published a paper on the telemetry of brain waves in the "Proceedings" of the International Telemetering Conference, 1972. Byrd was a physical scientist at the Naval Surface Weapons Center, White Oaks Laboratory, Silver Springs, Maryland (1968- unknown, at least 1981) Byrd describes his work with Naval Surface Weapons as "predicting what war will be like in the future."

B. Dr. Gordon J. F. MacDonald - science advisor to President Lyndon Johnson - wrote, "Perturbation of the environment can produce changes in behavioral patterns." He was referring to low-frequency EM waves in the ionosphere affecting human brain wave patterns. (From his book, Unless Peace Comes, a Scientific Forecast of New Weapons, cited in "New World Order ELF Psychotronic Tyranny", a paper by C. B. Baker.)

C. SPS: Solar Power Satellite Project

In 1968, the U.S. military proposed Solar Powered Satellites in geostationary orbit some 40,000 km above the Earth, which would intercept solar radiation using solar cells on satellites and transmit it via a microwave beam to receiving antennas (called rectennas) on Earth.

The U.S. Congress mandated the Department of Energy and NASA to prepare an Environmental Impact Assessment on this project, to be completed by June 1980 and costing $25 million. This project was designed to construct 60 Solar Powered Satellites over a 30-year period at a cost between $500 and $800 thousand million (in 1968 dollars), providing 100 percent of the US energy needs in the year 2025 at a cost of $3000 per kW.

At that time, the project cost was 2-to-3 times larger than the whole Department of Energy budget and the projected cost of the electricity was well above the cost of most conventional energy sources. The rectenna sites on Earth were expected to take up to 145 square kilometers of land and would preclude habitation by any humans, animals, or even vegetation.

Each Satellite was to be the size of Manhattan Island. [note: Sounds curiously like the HAARP array, yes?]

D. CIA experiments with the possibility of poisoning drinking water by injecting chemicals into the water supply of the FDA in Washington, DC.

1969

A. Charles Tart studied electrical engineering at MIT and received a PhD in psychology from the University of North Carolina. He taught humanistic and experimental psychology at the University of California, Davis. Has served as Instructor in Psychiatry at the University of Virginia Medical School, and as Lecturer in Psychology at Stanford University.

His work has dealt with parapsychology, sleep and dreaming, hypnosis, and psychoactive drugs. [Tart, Charles, ed., Altered States of Consciousness, Anchor Books, 1969, inside cover]

B. Dr. Robert MacMahan of the Department of Defense requests from Congress $10 million to develop - within 5-to-10 years - a synthetic biological agent to which no natural immunity exists.

1970

A. Zbigniew Brzezinski - President Jimmy Carter's National Security Director - said in his book Between Two Ages that weather control was a new weapon that would be the key element of strategy. "Technology will make available to leaders of major nations a variety of techniques for conducting secret warfare..."

He also wrote that "Accurately-timed, artificially-excited electronic strokes could lead to a pattern of oscillations that produce relatively high-power levels over certain regions of the Earth ... One could develop a system that would seriously impair the brain performance of a very large population in selected regions over an extended period." [Cited in Baker's "ELF Psychotronic Tyranny" paper.]

B. Funding for the synthetic biological agent is obtained under H.R. 15090. The project - under the supervision of the CIA - is carried out by the Special Operations Division at Fort Detrick, the Army's top-secret biological weapons facility. Speculation is raised that molecular biology techniques are used to produce AIDS-like retroviruses.

C. United States intensifies its development of "ethnic weapons" (Military Review, Nov., 1970), designed to selectively target and eliminate specific ethnic groups who are susceptible due to genetic differences and variations in DNA.

1971

A. Hal Puthoff joined SRI in 1971 as a specialist in laser physics.

B. circa 1972- Hubbard was hired by Willia Harman, (then director of the Educational Policy Research Center at SRI to be a special investigative agent) earning $100 a day. Officially he was a security guard although his actual duties included spying on the drug culture which Hubbard - a political conservative - disdained. He stayed at SRI until the late 1970s (Lee and Schlain, pg. 198-9).

C. According to Jack Sarfatti, a "very, very sophisticated and successful covert psychological warfare operation run by the late Brendan O Regan of the Institute of Noetic Sciences and the late Harold Chipman who was the CIA station chief responsible for all mind-control research in the Bay Area in the 70s."

1972

A. Bruce Maccabee: Dr. Maccabee has been a Research Physicist at the Naval Surface Weapons Center in Silver Spring, Maryland since 1972. His work has centered on high power lasers, underwater sound, and the Ballistic Missile Defense. He holds a Ph.D. in Physics from the American University in Washington, DC. Dr. Maccabee was a member of the National Investigations Committee on Aerial Phenomena (i.e., UFOs).

B. In early 1972, psychic Ingo Swann heard of Hal Puthoff's research proposal through Cleve Backster. according to Swann, Backster maintained his intelligence connections, and Backster reported that the CIA was interested in his experiments. Some of Backster's experiments are documented in "PRIMARY PERCEPTION: Cleve Backster's astounding mind/plant communication discovery!", Australian Lateral Thinking Newsletter,1996.

C. Puthoff is head of the SRI remote-viewing program, 1972-85. After he left, Puthoff was replaced with Ed May, a former Naval Intelligence Officer. (Puthoff, Harold, "CIA-Initiated Remote Viewing Program at Stanford Research Institute", Journal of Scientific Exploration, Vol. 10, No. 1, Spring 1996)

D. The Taser - the first electrical shock device developed for use by law enforcement - delivers barbed, dart-shaped electrodes to a subject's body and 50,000 volt pulses at 2-millionths of an amp over 12-14 seconds time. (From "Report on the Attorney General's Conference on Less Than Lethal Weapons", by Sherry Sweetman, 1987, p 4, which cites "Non-Lethal Weapons for Law Enforcement: Research Needs and Priorities. A Report to the National Science Foundation by the Security Planning Corporation, 1972. Research by Harlan Girard.)

E. "A U.S. Department of Defense document said that the Army has tested a microwave weapon. It was an extremely powerful 'electronic flamethrower'. " (From Electromagnetic Pollution)

F. "A study published by the U.S. Army Mobility Equipment Research and Development Center, titled 'Analysis of Microwaves for Barrier Warfare' examines the plausibility of using radio frequency energy in barrier counter-barrier warfare

The report concludes that

a. it is possible to field a truck-portable microwave barrier system that will completely immobilize personnel in the open with present day technology;

b. there is a strong potential for a microwave system that would be capable of delaying or immobilizing personnel in vehicles;

c. with present technology, no method could be identified for a microwave system to destroy the type of armored material common to tanks." (From Electromagnetic

Pollution by Kim Besly, p 15, quoting The Zapping of America by Paul Brodeur.) The report further documents the ability to create third-degree burns on human skin using 3 Gigahertz at 20 watts/square-centimeter in 2 seconds.

G. Dr. Gordon J. F. MacDonald testified before the House Subcommittee on Oceans and International Environment concerning low-frequency research: "The basic notion there was to create between the electrically-charged ionosphere in the higher part of the atmosphere and conducting layers of the surface of the Earth this neutral cavity, to create waves - electrical waves that would be tuned to the brainwaves ... about 10 cycles per second ... you can produce changes in behavioral patterns or in responses." [from Baker's "ELF Psychotronic Tyranny" paper.]

1973

A. Sharp and Grove transmit audible words via microwaves [EW: That is, voice to SKULL] (See "Synthetic Telepathy" in Resonance]

B. "Richard Kennett" is a pseudonym used by author Jim Schnabel in Remote Viewers (Dell, 1997) to describe a CIA scientist who worked with the remote-viewing project. In the photo insert is a picture of Kennett, Pat Price, and Harold Puthoff after a remote-viewing experiment involving a glider.

Elsewhere (example: Puthoff, Harold, "CIA Remote Viewing Program at Stanford" ", Journal of Scientific Exploration, Vol. 10, No. 1, Spring 1996), the man in the photo on the left is identified as Chistopher Green. As there can't be too many scientists at the CIA with an interest in the paranormal with this name, I feel safe in guessing that the two is the same, although I haven't absolutely confirmed it. At any rate, here is the information on "Richard Kennett", all from Remote Viewers. In Spring 1973, he was an analyst with the

CIA's Office of Scientific Intelligence with a Ph.D. in neurophysiology.

"Within a decade, Kennett would be the assistant national intelligence officer for chemical and biological warfare issues". His work concentrated on evaluating the health of foreign officials, but he also explored the fringes of medicine and psychology. It was under these circumstances that he challenged Hal Puthoff's research at SRI, although he was not officially controlling the contract. (pg. 104-6) The initial challenge was to view a secret microwave receiving station. According to Schnabel's information, this would make Kennett the "East Coast challenger" from Mind Reach]. Kennett - as well as the team at SRI - were reportedly investigated by the Defense Investigative Service after the viewing. Kennett was also involved with the experiments with Uri Geller. (pg. 139).

Kennett was also called in to look at the scientists at the Lawrence Livermore National Laboratory who began to see "visions" after experimenting with Geller. (pg. 166-9) Kennett left the CIA around 1985. (pg. 317)

1974

A. Monroe Institute. Founded and directed by Robert Monroe from 1974 until his death in1995. Had classified contracts with the U.S. Army Intelligence & Security Command (INSCOM) on orders by Gen. Albert Stubblebine. The Institute studied their hemi-synch techniques to see if they could enhance soldiers' performance and concentration. (Emerson, Steven, Secret Warriors, G.P. Putnam's Sons, 1988, pg. 103-4) The primary area of research at the Monroe Institute involves using a binaural beat to cause different psychological effects. A binaural beat is created by using stereo headphones with each speaker emitting a slightly different frequency. The result is a tone at the frequency between the two, which allegedly causes the brain to "entrain" on the frequency (i.e. the brain waves regulate themselves to the same frequency).

The National Research council evaluated the Institute's claims that the method could be used to improve learning. [National Research Council, Enhancing Human Performance, National Academy of Sciences, 1988, pg. 111-4]

"located near Charlottesville, Virginia. Bob Monroe - author of many books on 'Out of Body Experiences' - has long and close ties with the CIA. James Monroe - Bob's father, if I'm not mistaken - was involved with the Human Ecology Society - a CIA front organization of the late 50s and 60s.

The Monroe Institute has done research on accelerated learning and foreign language learning through the use of altered states of consciousness for the CIA and other government organizations.

Government interest in the more radical research going on at the Institute remains only tantalizing speculation. Official classified document storage boxes have been seen at their mail-order outlet located in Lovingston, VA." (Porter, Tom, Government Research into ESP & Mind Control, March, 1996)

The Monroe Institute trained the government viewers from Ft. Meade in Out of Body Experiences (OBEs). Courtney also went through this training which involves using the Institute's Hemisync tapes. These tapes - which work by using a binaural beat to entrain brain waves - caused Brown to feel that he left his body and communicated with aliens. [Brown, Courtney, Cosmic Voyage, Dutton, 1996]

B. In 1974, Jack Sarfatti is director of a physics program at the Esalen Institute. He's been funded by Werner and Jean Lanier (a friend of Laurance Rockefeller). (Sarfatti, Jack, "The Parsifal Effect", The Destiny Matrix) Sarfatti met with Puharich, Uri Geller, and Ira Einhorn at Puharich's Ossining ranch. Einhorn acted as a literary agent for Sarfatti and brought him to Esalen Physics /Consciousness research group. This is where it all started back in 1975. PCRG was co-founded by Jack Sarfatti and Michael Murphy at the Esalen Institute in Big Sur,

California in 1974. Financed by Werner Erhard, Jean Lanier, and the late George Koopman, the PCRG nurtured the creation of books like Space-Time and Beyond, The Tao of Physics, The Dancing Wu Li Masters, Cosmic Trigger, and The Roots of Consciousness.

The group included the physicists and authors Fred Alan Wolf, Nick Herbert, and Fritjof Capra, along with Saul Paul Sirag, Henry Dakin, Robert Anton Wilson, Uri Geller, Barbara Honneger, the late Brendan O Regan, George Leonard, Gary Zukav, Ira Einhorn, and artist Lynn Hershmann. Nobel Laureate Brian Josephson along with physicists David Finkelstein, Russell Targ, Karl Pribram, Henry Stapp, Phillipe Eberhard, and Ralph Abraham all came for shorter visits. The group is now reborn on the World Wide Web 20 years later with both new and old faces. According to George Koopman, the PCRG was the inspiration for the film Ghost.

1975

A. Saturn V Rocket. Due to a malfunction, the Saturn V rocket burned unusually high in the atmosphere - above 300 km. This burn produced "a large ionospheric hole" (Mendillo, M. et al., Science, p. 187, 343, 1975). The disturbance reduced the total electron content more than 60% over an area 1,000 km in radius and lasted for several hours. It prevented all telecommunications over a large area of the Atlantic Ocean.

The phenomenon was apparently caused by a reaction between the exhaust gases and ionospheric oxygen ions. The reaction emitted a 6300 Å airglow. Between 1975 and 1981, NASA and the U.S. Military began to design ways to test this new phenomenon through deliberate experimentation with the ionosphere.

B. Bruce Maccabee joined MUFON and was appointed State Director for Maryland and a Consultant in Photo Analysis and Laser Physics.

C. In the 1970s, Mike Murphy became interested in Russian parapsychology and visited the country to meet experimenters in this field. This led to a close connection between Esalen and some Russian officials, who set up an exchange program. Lasting into the 1980s, this exchange was dubbed "hot-tub diplomacy". John Mack was reportedly involved in this exchange. In the late 1970s, Esalen became involved with an Englishwoman named Jenny O'Connor, who claimed to be in psychic contact with 'the Nine' (probably the same "Nine" that Andriaj Puharich claimed to be in contact with). Dick Price and other members of the Esalen staff became increasingly dependent on 'the Nine' to the point of listing them as program leaders and members of the Esalen Gestalt Staff in brochures. (Anderson, pg. 302)

D. The virus section of Fort Detrick's Center for Biological Warfare Research is renamed the Fredrick Cancer Research Facilities and placed under the supervision of the National Cancer Institute (NCI). It is here that a special virus cancer program is initiated by the U.S. Navy, purportedly to develop cancer-causing viruses. It is also here that retrovirologists isolate a virus to which no immunity exists. It is later named HTLV (Human T-cell Leukemia Virus).

1975-1977

"Unpublished analyses of microwave bioeffects literature were disseminated to the U.S. Congress and to other officials arguing the case for remote control of human behavior by radar." (From the Journal of Microwave Power, 12(4), 1977, p 320. Research by Harlan Girard.)

1976

A. Around late-1976 to 1977, Dale Graff - then a physicist with the Air Force's Foreign Technology Division - gave a small contract to the SRI research team. Graff wanted to replicate some Soviet psi experiments done in submarines, as well as test the Soviet hypothesis that psi was transmitted via ELF (Extremely Low Frequency) electromagnetic waves.

These tests were conducted in July, 1977 with the help of Stephan Schwartz, a former Navy officer and psychic researcher. Schwartz helped procure a submarine for a July 1977 experiment with SRI. These experiments included some on behalf of Dale Graff of the Air Force. (Schnabel, Jim, Remote Viewers: The Secret History of America's Psychic Spies, Dell, 1997, pg. 207) Research associate with the Cognitive Sciences Laboratory.

B. Around 1976, Koslov - as the scientific assistant to the secretary of the Navy - was being briefed on various contracts the Navy held, including one for SRI. The section describing the contract at SRI was headed "ELF AND MIND CONTROL" (ELF stands for Extremely-Low Frequency). Reportedly, Koslov was upset by the label and cancelled the contract with SRI. "I don't believe it's the function of the military to support parapsychology." (Wilhelm, John, "Psychic Spying?", Washington Post 08/07/77, B5) According to another account, the heading was "Sensing of Remote EM sources (Physiological Correlates)". According to this account, Koslov thought the project dealt with mind-control and looked into the contract in more detail. He found that it dealt with psychic research which upset him as well, and ordered the contract to be cancelled. (Schnabel, Jim, Remote Viewers, Dell, 1997, pg. 206) Either Wilhelm paraphrased and misinterpreted the section heading on the briefing, or the story was sanitized somewhere along the line before reaching Schnabel's book. In either case, the Navy continued to fund psychic research (Wilhelm, 1977) and has been one of the biggest funders of research related to electronic mind-control.

1977

A. Christopher Bird presented a paper on "dowsing" and the psychic ability of plants at the "Mind Over Matter" conference at Penn State University, late January,\ 1977, organized by Ira Einhorn. Other attendees included Andrija Puharich and Thomas Bearden (Levy, pg. 189).

B. Soon afterwards, Einhorn and the "Psychic Mafia" focused their attention on ELF mind-control (Levy, pg. 190). He suggests that his murder charge could have been a set-up by the CIA or KGB for his interest in activities by America and Russia in the areas of psychic warfare, Tesla technology, and mind-control (Levy, pg. 242). Puharich says that Einhorn's work wasn't important enough to elicit such a reaction (Levy, pg. 308). The likelihood is that Einhorn - like many of the individuals involved in COINTELPRO - was merely a "useful idiot" who was as manipulated as those he sought to manipulate. Einhorn led seminars at the Esalen Institute and was involved with the Physics/Consciousness Research Group. He reportedly worked with Congressman Charlie Rose, a large supporter of psychic studies, on classified projects. Senate hearings on Health and Scientific Research confirm that 239 populated areas had been contaminated with biological agents between 1949 and 1969. Some of the areas included San Francisco, Washington, D.C., Key West, Panama City, Minneapolis, and St. Louis.

1978

A. Experimental Hepatitis-B vaccine trials conducted by the CDC begin in New York, Los Angeles, and San Francisco. Ads for research subjects specifically ask for promiscuous homosexual men.

B. Hungarians presented a state-of-the-art paper on infrasonic weapons to the United Nations, "Working Paper on Infrasound Weapons", United Nations CD/575, 14 Aug 1978 (from The Road from Armageddon by Peter Lewis).

C. SPS Military Implications

Early review of the Solar Powered Satellite project began in around 1978, and I [Rosalie Bertell] was on the review panel. Although this was proposed as an energy program, it had significant military implications. One of the most significant - first pointed out by Michael J. Ozeroff - was the possibility of developing a satellite-borne beam weapon for anti-ballistic missile (ABM) use.

The satellites were to be in geosynchronous orbits with each providing an excellent vantage point from which an entire hemisphere can be surveyed continuously. It was speculated that a high-energy laser beam could function as a thermal weapon to disable or destroy enemy missiles. There was some discussion of electron weapon beams through the use of a laser beam to preheat a path for the following electron beam. The SPS was also described as a psychological and anti-personnel weapon which could be directed toward an enemy. If the main microwave beam was redirected away from its rectenna toward enemy personnel, it could use an infrared radiation wavelength (invisible) as an anti-personnel weapon. It might also be possible to transmit high enough energy to ignite combustible materials.

Laser beam power relays could be made from the SPS satellite to other satellites or platforms - for example, aircraft - for military purposes. One application might be a laser powered turbofan engine which would receive the laser beam directly in its combustion chamber, producing the required high temperature gas for its cruising operation. This would allow unlimited on-station cruise time. As a psychological weapon, the SPS was capable of causing general panic. The SPS would be able to transmit power to remote military operations anywhere needed on Earth. The manned platform of the SPS would provide surveillance and early warning capability as well as ELF linkage to submarines. It would also provide the capability of jamming enemy communications. The potential for jamming and creating communications is significant. The SPS was also capable of causing physical changes in the ionosphere. President Carter approved the SPS project and gave it a go-ahead in spite of the reservation which many reviewers - myself included - expressed. Fortunately, it was so expensive - exceeding the entire Department of Energy budget - that funding was denied by the Congress. I approached the United Nations Committee on Disarmament on this project.

But I was told that as long as the program was called "solar energy" by the United States, it could not be considered a "weapons" project. The same project resurfaced in the US under President Reagan. He moved it to the much larger budget of the Department of Defense and called it "Star Wars". Since this is more recent history, I will not discuss the debate which raged over this phase of the plan. By 1978, it was apparent to the U.S. Military that communications in a nuclear hostile environment would not be possible using traditional methods of radio and television technology (Jane's Military Communications 1978).

By 1982, GTE Sylvania (Needham Heights, Massachusetts) had developed a command & control electronic sub-system for the U.S. Air Force's Ground-Launched Cruise Missiles (GLCM) that would enable military commanders to monitor and control the missile prior to launch both in hostile and non-hostile environments. The system contains 6 radio subsystems, created with visible light using a dark beam (not visible) and is resistant to the disruptions experienced by radio and television. Dark beams contribute to the formation of energetic plasma in the atmosphere. This plasma can become visible as smog or fog. Some has a different charge than the Sun's energy and accumulates in places where the Sun's energy is absent, like the polar regions in the Winter. When the polar Spring occurs, the Sun appears and repels this plasma, contributing to holes in the ozone layer.

This military system is called Ground Wave Emergency Network (GWEN). (See the SECOMII Communication

System, by Wayne Olsen, SAND 78-0391,Sandia Laboratories, Albuquerque, New Mexico, April 1978.)

This innovative emergency radio system was apparently never implemented in Europe and exists only in North America.

1979

A. In February 1979, Alfred Hubbard attended an LSD reunion party hosted by Dr. Oscar Janiger along with Laura Huxley, Sidney Cohn, John Lilly, Willis Harman, and Timothy Leary among others (Lee and Schlain, 213).

B. Around 1979, SRI funded a project of Tart's which screened university students and faculty for psychic ability. (Schnabel, Jim, Remote Viewers: The Secret History of America's Psychic Spies, Dell, 1997, pg. 225-6)

C. In an article entitled "The Fund for CIA Research, or Who's Disinforming Who?", the anonymous authors (the Associated Investigators Group) accuse Bruce Maccabee of working with the CIA, providing them with information, and letting the CIA affect his leadership in FUFOR. According to the article, Maccabee's main contact with the CIA was through Dr. Christopher Green. In a written response, Maccabee rebuts that most of his contacts with the CIA have been in the context of his work with the Navy and are unrelated to his UFO research. He says that he did give CIA employees informal lectures at the request of Ron Pandolfi, but that the CIA has never attempted to influence his research.

[A similar rebuttal was written by Aviary guy Dan Smith and Rosemary Ellen Guiley of Fate Magazine, and New Age Land Central - in later years - after similar accusations were made.]

"I never contacted any companies. What I did was tell Jack Acuff - Director of NICAP at the time - that I would like to speak to experts in the field of radar. He, in turn, put me in contact with a scientist - Dr. Gordon MacDonald - at the MITRE corporation. I was invited to discuss the NZ sightings with him and several other scientists at MITRE in McLean, VA.

And I did (and they generally agreed with my conclusions). Then a week-or-so later, I learned that MacDonald had contacted a man at the CIA who contacted me and offered to provide technical consultation if I would provide a briefing to some CIA employees. At first, I was leery of doing anything with the CIA. But I knew they had radar experts, so I stipulated that if they would give me some feedback I'd tell them what I know.

So, I briefed them and I received some helpful comments..." [note: When you dance with the Devil, the Devil doesn't change - the Devil changes YOU!] "After I discussed the NZ case one employee - Dr. Christopher "Kit" Green (KG) - invited me to visit the CIA again a week-or-so later to have a general UFO discussion with him and a couple of other employees... After that last meeting with KG in the spring of 1979, I didn't see him again and had no contact with the agency until June, 1984 when I was contacted by Dr. Ronald Pandolfi regarding my Navy work.

He had been tracking developments by the "other side" in that field of research and wanted to know what the U.S. state-of-the-art was." (Bruce Maccabee's response to the AIR report)

Formerly with the CIA, Dr. Green's work involved UFO research.

"Dr. Green attained a Ph.D. in Neurophysiology in 1969 and in 1976 received his M.D. (Doctor of Medicine) degree. Green was awarded the CIA's National Intelligence Medal for his work on a 'classified project' from 1979 to 1983 - precisely the years in which Maccabee was meeting with him at CIA headquarters.

Green uses somewhat of a cover story to describe his CIA work, calling himself a 'Scientific Advisor on the Advisory Board to the Directorate of Intelligence, CIA.'" (The Associated

Investigators Group, "The Fund for CIA Research, or Who's Disinforming Who?")

Esalen also held seminars in quantum physics, and was the birthplace of the Physics/Consciousness Research Group. Some results of these seminars are documented in Zukav, Gary, The Dancing Wu Li Masters, Morrow Quill, 1979

1980

A. By the 1980's, Koslov was working with the Applied Physics Laboratory at Johns Hopkins University, where he continued to study the effects of electromagnetic radiation on humans. He is currently the vice president of the Maryland Microscopical and Scientific Instrument Society.

B. Dale Graff had continued to task SRI on behalf of the Air Force for the next few years. In 1980, he won a fellowship for "exceptional analyst" within the intelligence community and planned to take 2 years off to conduct research in other laboratories: SRI, a psychokenesis lab at Princeton, a J.B. Rhine affiliated lab in Durham, NC, and a Department of Energy lab where microwave weapons were being studied. His fellowship was revoked by the office of the Air Force Chief of Staff and - with the encouragement of Jack Vorona - he retired from the Air Force and moved to the DIA, where he ran the Advanced Concepts Office.

C. "Michelle Smith" and Lawrence Pazder published "Michelle Remembers" about Satanic Ritual Abuse memories. She came to therapist Pazder because she was in distress over horrible dreams and a miscarriage.

1981

A. Orbit Maneuvering System

Part of the plan to build the SPS space platforms was the demand for reusable space shuttles since they could not afford to keep discarding rockets.

In 1981, The NASA Spacelab-3 mission of the Space Shuttle made "a series of passes over a network of 5 ground based observatories" in order to study what happened to the ionosphere when the Shuttle injected gases into it from the Orbit Maneuvering System (OMS). They discovered that they could "induce ionospheric holes" and began to experiment with holes made in the daytime or at night over Millstone, Connecticut and Arecibo, Puerto Rico.

They experimented with the effects of,

"artificially induced ionospheric depletions on very low frequency wave lengths, on equatorial plasma instabilities, and on low frequency radio astronomical observations over Roberval, Quebec, Kwajelein, in the Marshall Islands, and Hobart, Tasmania" (Advanced Space Research, Vol.8, No. 1, 1988).

B. Eldon Byrd- who worked for Naval Surface Weapons, Office of Non-Lethal Weapons - was commissioned in 1981 to develop electromagnetic devices for purposes including 'riot control', clandestine operations and hostage removal.

"Byrd also wrote of experiments where behavior of animals was controlled by exposure to weak electromagnetic fields. 'At a certain frequency and power intensity, they could make the animal purr, lay down, and roll over.'" (Keeler, Anna, "Remote Mind Control Technology") "Between 1981 and September 1982, the Navy commissioned me to investigate the potential of developing electromagnetic devices that could be used as non-lethal weapons by the Marine Corp for the purpose of 'riot control', hostage removal, clandestine operations, and so on." Eldon Byrd, Naval Surface Weapons Center, Silver Spring MD. (from "Electromagnetic Pollution" by Kim Besly, p 12.)

C. John Alexander supported the views of Thomas Bearden. Delivered a paper to the 1981 national convention of the US Psychotronic Association

D. General Albert Stubblebine. Former head of the U.S. Army Intelligence & Security Command (INSCOM) 1981-84, Master's degree in chemical engineering from Columbia. He signed classified contracts with the Monroe Institute (Emerson, Steven, Secret Warriors, G.P. Putnam's Sons, 1988, pg. 103-4). Stubblebine often met with Noriega while he was a U.S. intelligence asset (Emerson, 1988, pg. 110-1).

Stubblebine was the former boss of Col. John Alexander, and the two have held numerous "spoon-bending" parties. He is a friend of Lyn Buchanan [according to a representative from PSI TECH, the two are not friends]. Stubblebine is married to UFOlogist Rima Laibow. (Porter, Tom, Government Research into ESP & Mind Control, March, 1996). Soon after becoming head of INSCOM, Stubblebine began a program called the "High Performance Task Force" - a series of methods to improve his officers' performance. These ranged from the neuro-linquistic programming of Tony Robbins to the hemisynch tapes of the Monroe Institute where Stubblebine often sent his officers.

(Schnabel, Jim, Remote Viewers: The Secret History of America's Psychic Spies, Dell, 1997, pg. 276) Following an incident involving an officer having a psychotic episode at the Monroe Institute, Stubblebine resigned in 1984. He was replaced by Major General Harry Soyster (Schnabel, 1997, pg. 316), formerly vice-president for 'Intelligence Systems' of BDM of McClean, Virginia. As of 1992, Chairman of PSI-TECH. "Laibow, Stubblebine, and UFOlogist Victoria Lacas (with [C.B. Scott] Jones in the shadows) toured Europe and the Soviet Union, where they have established a prodigious UFO/Psi network." (Durant, Robert J., "Will the Real Scott Jones Please Stand Up?") Stubblebine gave a lecture at the International Symposium on UFO Research - sponsored by the International Association for New Science - in Denver, Colorado (May 22-25, 1992).

This gives a good example of Stubblebine's coherence (or lack thereof) and paranoia (he often threatened to destroy the tape). Stubblebine claimed that none of the members of the remote-viewing program had prior psychic abilities or interests (but all other sources state that they did).

E. In the Summer of 1981, Pat Delgado brought to the attention of the national Press the existence of mysterious circular depressions in the fields at Cheesefoot Head, Hampshire.

F. Budd Hopkins published Missing Time with an afterward by therapist Aphrodite Clamar. Hopkins book was about the in-depth investigation of 19 cases of UFO abduction which he had undertaken in the previous 5 years. (G) The first cases of AIDS are confirmed in homosexual men in New York, Los Angeles, and San Francisco, triggering speculation that AIDS may have been introduced via the Hepatitis-B vaccine.

1982

A. In May 1982, Elisabeth and Russell Targ held a workshop on psychic phenomena for 25 professionals.

This was part of a program with Stanislav Grof, who was studying non-chemical alternatives for altered states of consciousness. The Targs goal was to show that psychic experiences did not require an "altered state". (Targ, Russell and Harary, Keith, Mind Race, Villard Books, 1984, pg. 99). Grof served briefly as the branch chief of the operational unit of Star Gate from around 1982 or 83 until he resigned in summer of 1993.

B. Electromagnetic weapons for law enforcement use in Great Britain.

A 10-30 Hz strobe light which can produce seizures, giddiness, nausea, and fainting was developed by Charles Bovill of the now defunct British firm Allen International. Addition of sound pulses in the 4.0-7.5 Hz range increases effectiveness as utilized in the Valkyrie - a "frequency" weapon advertised in British Defense Equipment Catalogue until 1983.

The squawk box or "sound curdler" uses 2 loudspeakers of 350-watt output to emit 2 slightly different frequencies which combine in the ear to produce a shrill shrieking noise. The U.S. National Science Foundation report says there is "severe risk of permanent impairment of hearing." (From Electro pollution by Kim Besley, citing the Manchester City Council Police Monitoring Unit document.)

C. Air Force review of biotechnology.

"Currently available data allow the projection that specially generated radiofrequency radiation (RFR) fields may pose powerful and revolutionary antipersonnel military threats. Electroshock therapy indicates the ability of induced electric current to completely interrupt mental functioning for short periods of time to obtain cognition for longer periods and to restructure emotional response over prolonged intervals. "... impressed electromagnetic fields can be disruptive to purposeful behavior and may be capable of directing and/or interrogating such behavior. Further, the passage of approximately 100 milliamperes through the myocardium can lead to cardiac standstill and death, again pointing to a speed-of-light weapons effect. "A rapidly scanning RFR system could provide an effective stun or kill capability over a large area." (From "Final Report on Biotechnology Research Requirements for Aeronautical Systems Through the Year 2000". AFOSR-TR-82-0643, Vol 1, and Vol 2, July 30, 1982.)

1983

Phoenix II / USAF, NSA:

Location: Montauk, Long Island

Electronic multi-directional targeting of select population groups

Targeting: Medium range

Frequencies: Radar, microwaves. EHF UHF modulated

Power: Gigawatt through Terawatt

Purpose: Loading of Earth Grids, planetary sonombulescence to stave off geological activity, specific-point earthquake creation, population programming for "sensitized" individuals

Pseudonym: "Rainbow", ZAP

A. Nikolai Khokhlov - a Soviet KGB agent who defected to the West in 1976 - interviews recently arrived scientists and reports that "The Soviet mind-control program is run by the KGB with unlimited funds." (From the Spectator, Feb 5, 1983, reported in "New World Order Psychotronic Tyranny" by C. B. Baker.)

B. "Center Lane" was the codename for the operational unit of the remote-viewing program, redesignated from Grill Flame in late 1983. Control of the unit shifted from INSCOM's operation group to the more direct control of Albert Stubblebine. The unit was known as INSCOM Center Lane Project (ICLP). (Schnabel, Jim, Remote Viewers: The Secret History of America's Psychic Spies, Dell, 1997, pg. 280) In late 1983, 4 more individuals were recruited to Center Lane: Captain Ed Dames, Captain Bill Ray (counterintelligence specialist), Captain Paul Smith, and Charlene Cavanaugh (civilian analyst with INSCOM). These four began a training program - which started at The Monroe Institute - and concluded with personal training with Ingo Swann. (Schnabel, 1997, pg. 292-3)

After Gen. Stubblebine's retirement in 1984, Center Lane was completely without support in the Army. Jack Vorona arranged for the unit to be transferred directly to the DIA's Scientific and Technical Intelligence Directorate when Army funding ran out in late 1985, at which time it was redesignated Sun Streak. Until that time, the unit was given no official tasking (Schnabel, 1997, pg. 319). Center Lane started when Ingo Swann at SRI came across a breakthrough in his techniques in 1983. He developed a training program and trained 6 military officers (including Ed Dames) over a period of 6 months. After finishing the training in late 1983, the viewers returned and started applying their knowledge.

The unit was renamed 'Center Lane' with Dames as the operations and training officer. "Dames took a 'let's see what this baby can do' approach, replacing the unit's former intelligence collection methodology with the breakthrough technique." (Dames, Ed, "Ed Dames Sets the Record Straight") [Keep in mind that Dames is a major disinfo artist.]

1984

"USSR: New Beam Energy Possible?", possibly associated with early Soviet weather engineering efforts over the U.S. (from "Tesla's Electromagnetics and Its Soviet Weaponization" by T.E. Bearden.) According to former Reagan aide Barbara Honneger,

"the fundamental reason for the increased interest [in psi research] is initial results coming out of laboratories in the United States and Canada that certain amplitude and frequency combinations of external electromagnetic radiation in the brain-wave frequency range are capable of bypassing the external sensory mechanism of organisms - including humans - and directly stimulating higher-level neuronal structures in the brain.

This electronic stimulation is known to produce mental changes at a distance, including hallucinations in various sensory modalities - particularly auditory." (McRae, Ronald, Mind Wars, St. Martin's Press, 1984, pg. 136)

The overlap between these 2 fields can be described as:

Mis-identification:

Some ELF mind-control studies have been discussed under the heading of "psychotronics". Many - myself included - don't agree with this label as there is no psychic component in the study of the effects of electromagnetic radiation on the central nervous system.

Coincidental Findings:

As in most scientific fields, research that is tangential for one project may be central to another. Navy studies in ELF communications included a portion on possible health effects. When these findings were revealed, the possibility of using ELF as a weapon arose and studies were continued in that direction. However, we can't say that all of the Navy's research into ELF radio was a front for mind-control as they have a definite interest in communication with their submarines. The same may be true for remote-viewing studies. Studies at SRI and elsewhere measured and analyzed subject's brain waves, and also studied the effects of ELF waves as a possible carrier for telepathic information. T

Tech-Enhanced Psi:

Some studies - especially those involving dolphins - tried to use technology to enhance psychic phenomena. Most of this is pure bunk including most of the inventions I've seen created by the Russians and the US Psychotronic Association. Some of it resembles telepathy simulated by technology, such as the attempt to carry signals from dolphins to humans via the "Neurophone".

This would seem to fit better under "Mis-identification".

Cover:

Remote-viewing - like UFOs - has been postulated by some researchers as being used as a "cover story" for covert mind-control experiments. This plan would convince the victims that the "voices" or sensory data they were unnaturally receiving was due to channeling, telepathy, or remote-viewing. It would also have the "high-weirdness" factor, which would preclude a serious treatment of the subject by the mainstream media. However, I'm hesitant to lump the entire spectrum of government interest in psi in this category.

Cutting Edge:

Both psychic ability and things like non-lethal weapons are considered to be on the "cutting edge" of military theory. This is an alternative explanation as to why individuals like John Alexander and David Morehouse are interested in both fields. The degree to which these crossovers apply to specific cases are dealt with individually, and to this subject as a whole in the conclusions.

1985

A. Founded by Ed May, the Cognitive Sciences Laboratory was formed at SRI in 1985 and moved with May to SAIC. May and the Cognitive Sciences Laboratory are currently at a "small start-up research place called the Laboratories for Fundamental Research" (e-mail from Ed May, 8/7/96).

Joe McMoneagle is listed as a research associate. Other staff members include S. James, P. Spottiswoode, Earling DeGraff, Nevin D. Lantz, Philip Wasserman, Laura V. Faith, Ellen Messer, and Stephan A. Schwartz.

"I (Dean Radin) took a leave of absence from Bell Labs in 1985 and spent that entire year at SRI International, working with Hal Puthoff and Ed May.

Since then, I spent about half my time in academia (Princeton, Edinburgh, UNLV) and half in industry (Contel Technology Center, GTE Labs). My academic research was exclusively on psi phenomena, and my industrial research included about 20% on psi. "I'm not in favor of developing or using psi for any military purposes. But unfortunately, there are those in the World who would use psi as a weapon if they could.

Thus, I reluctantly suppose that R& D on psi for intelligence and possibly military purposes can be justified for defensive reasons. It would be naive to think that someone, somewhere is not working on this right now."

(Interview with the Retro Psychokinesis Project)

B. Since the early 1970s, Puthoff had been a part-time paid consultant to Bill Church regarding alternative fuel sources. At Puthoff's urging, Church developed a company (Jupiter Technologies) to research Zero-Point Energy. In the summer of 1985 after giving only 2 week's notice, Puthoff left SRI to work for Church full time. (Schnabel, Jim, 1997, pg. 323)

C. Women in the peace camps at Greenham Common began showing various medical symptoms believed to be caused by EM surveillance weapons beamed at them. (See "Zapping: The New Weapon of the Patriarchy", Resonance#13, pp 22-24. Research by Woody Blue.)

D. Innovative Shuttle Experiments An innovative use of the Space Shuttle to perform space physics experiments in Earth orbit was launched, using the OMS injections of gases to "cause a sudden depletion in the local plasma concentration - the creation of a so-called ionospheric hole". This artificially-induced plasma depletion can then be used to investigate other space phenomena, such as the growth of the plasma instabilities or the modification of radio propagation paths.

The 47 second OMS burn of July 29, 1985 produced the largest and most long-lived ionospheric hole to date, dumping some 830 kg of exhaust into the ionosphere at sunset. A 6-second, 68-km OMS release above Connecticut in August 1985 produced an airglow which covered over 400,000 square km. During the 1980s, rocket launches globally numbered about 500-to-600 a year, peaking at 1500 in 1989. There were many more during the Gulf War. The Shuttle is the largest of the solid fuel rockets with twin 45-meter boosters. All solid fuel rockets release large amounts of hydrochloric acid in their exhaust.

Each Shuttle flight injects about 75 tons of ozone-destroying chlorine into the stratosphere. Those launched since 1992 inject even more ozone-destroying chlorine (about 187 tons) into the stratosphere (which contains the ozone layer).

E. Whitley Strieber publishes Communion.

F. According to the journal Science (227:173-177), HTLV and VISNA - a fatal sheep virus - are very similar, indicating a close taxonomic and evolutionary relationship.

1986

A. Attorney General's Conference on Less Than Lethal Weapons

Reviews current weapons available. Most date back to 1972: the Taser; the Nova XR-5000 Stun Gun (can interrupt a pacemaker); the Talon, a glove with an electrical pulse generator; and the Source, a flashlight with electrodes at the base.

These devices are useful only at close range except for the Taser, and are generally restricted to correctional institutions. Photic driving strobe lights tested by one conference delegate on 100 subjects produced discomfort. Closed eyelids do not block the effect. Evidence that ELF produces nausea and disorientation. Suggestion to develop fast-acting electro sleep-inducing EM weapon. Discusses problem of testing weapons on animals and human "volunteers".

(From "Report on the Attorney General's Conference on Less Than Lethal Weapons", by Sherry Sweetman, March 1987, prepared for the National Institute of Justice. Research by Harlan Girard.)

B. "The Electromagnetic Spectrum in Low-Intensity Conflict" by Captain Paul Tyler, MC, USN quotes the above passage and further elaborates on the theme. (Published in Low Intensity Conflict and Modern Technology, Lt. Col. David J. Dean,

USAF, ed., Air University Press, Maxwell AFB, AL. Research by Harlan Girard.)

C. On 02/10/86, Cleve Backster's lab was visited by National Research Council's Committee on Techniques for the Enhancement of Human Performance. The NRC was evaluating enhancement and parapsychological studies conducted for the Army. So, it is likely that Backster's research was involved with the Government. (National Research Council, Enhancing Human Performance, National Academy of Sciences, 1988, pg. 193-8)

D. Mighty Oaks

In April 1986, just before the Chernobyl disaster, the U.S. had a failed hydrogen test at the Nevada Test Site called "Mighty Oaks". This test - conducted far underground - consisted of a hydrogen bomb explosion in one chamber with a leaded steel door to the chamber (2 meters thick) closing within milliseconds of the blast.

The door was to allow only the first radioactive beam to escape into the "control room" in which expensive instrumentation was located. The radiation was to be captured as a weapon beam. The door failed to close as quickly as planned, causing the radioactive gases and debris to fill the control room and destroying millions of dollar's worth of equipment. The experiment was part of a program to develop X-ray and particle beam weapons.

The radioactive releases from Mighty Oaks were vented under a "licensed venting" and were likely responsible for many of the North American nuclear fallout reports in May 1986, which were attributed to the Chernobyl disaster.

E. According to the Proceedings of the National Academy of Sciences (83:4007-4011), HIV and VISNA are highly similar and share all structural elements except for a small segment which is nearly identical to HTLV. This leads to speculation that HTLV and VISNA may have been linked to produce a new retrovirus to which no natural immunity exists.

F. A report to Congress reveals that the U.S. Government's current generation of biological agents includes modified viruses, naturally occurring toxins, and agents that are altered through genetic engineering to change immunological character and prevent treatment by all existing vaccines.

1987

A. In 1987, Pandolfi invited UFOlogist Bruce Maccabee "to give a general lecture to [CIA] employees on UFOs and MJ-12". (Maccabee's response to AIR)

B. Department of Defense admits that despite a treaty banning research and development of biological agents, it continues to operate research facilities at 127 facilities and universities around the Nation.

1988

A. After retiring from the Army in 1988, John Alexander joined the Los Alamos National Laboratories and began working with Janet Morris, the Research Director of the U.S. Global Strategy Council (USGSC) chaired by Dr. Ray Cline (deceased) former Deputy Director of the CIA."

B. The Pentagon is ordered by courts to cease EMP tests at several locations due to a lawsuit filed by an environmental group. (From the Washington Post, May 15, 1988, see "US and Soviets Develop Death Ray", Resonance 11, p 10. Research by Remy Chevalier.)

C. Senator Claiborne Pell from Rhode Island. Member of the Council on Foreign Relations and the Club of Rome. Decorated by the Knights of Malta. Along with Charlie Rose, Pell is one of Washington's biggest supporters of psychic research. In1988, he introduced a bill to get government funding for the New Age group the National Committee on Human Resources (Al Gore was a co-sponsor). He is also on the advisory board of the International Association of Near-Death Studies and on the board of the Institute of Noetic Sciences and the Human Potential Foundation. For 7 years, Claiborne Pell employed C.B. Scott Jones as an aide (Gardner, Martin, "Clairborne Pell: The Senator from Outer Space", Skeptical Inquirer, March/April 1996). Chairman of the Foreign Relations Committee. Pell was a close friend of BCCI figure Clark Clifford. (Truell, Peter and Gurwin, Larry, False Profits, Houghton Mifflin Company, 1992, pg. 240)

1989

A. MUFON appointed C.B. Scott Jones as a Special Consultant in International Relations

B. TRIDENT/ ONR, NSA:

Electronic directed targeting of individuals or populations

Targeting: Large population groups assembled

Display: Black helicopters flying in triad formation of three

Power: 100,000 watts

Frequency: UHF

Purpose: Large group management and behavior control, riot control Allied Agencies: FEMA

Pseudonym: "Black Triad" A.E.M.C

C. Human Potential Foundation founder and president C.B. Scott Jones. Board members include Clark Sandground and Claiborne Pell. Received original funding from Laurance Rockefeller. Passes funds from Rockefeller to UFO abduction researcher John Mack. Worked with Dr. Igor Smirnov.

D. Michael Persinger feels that he is able to replicate alien abduction and other supernatural phenomena through the use of 3 solenoids (attached to a modified motorcycle helmet) passing a magnetic pulse through the frontal lobes of the brain. Solenoids are called "magnetic coils" by psychiatrists, who use them as a non-intrusive alternative to implantable electrodes for stimulating the brain. (see Hallett, Mark and Cohen, Leonardo, "Magnetism: A New Method for Stimulation of Nerve and Brain", JAMA, 7/28/89, pg. 530)

E. John Alexander:

"I have served as chief of Advanced Human Technology for the Army Intelligence and Security Command (1982-84) and - during the preparation of the EHP [Enhancing Human Performance] report - was director of the Advanced Systems Concepts Office at the U.S. Army Laboratory Command."

Alexander stated that "psychotronic weapons lack traditional scientific documentation, and I do not suggest that research projects be carried out in that field." (Alexander, Col. John, "A Challenge to the Report", New Realities, March/April 1989)

F. Psi Tech founded in 1989 by president Ed Dames. Their vice-president is Jonina Dourif. A "John L. Turner" is listed as a monitor. Board Members include John B. Alexander and Gen. Albert Stubblebine.

1990

A. According to an anonymous BBC television reporter, Dr. Louis West headed up the medical oversight for the Ft. Meade remote-viewing operational unit. (Constantine, Alex, "'Remote Viewing' at Stanford Research Institute or Illicit CIA Mind Control Experimentation?") West was also a "Member of the medical oversight board for Science Applications International Corp. remote-viewing research in early 1990s.

RF MEDIA/ 1990, CIA:

Electronic, multi-directional subliminal suggestion and programming

Location: Boulder, Colorado (Location of main cell telephone node, national television synchronization node)

Targeting: national population of the United States

Frequencies: ULF VHF HF Phase modulation

Power: Gigawatts

Implementation: Television and radio communications, the "videodrome" signals

Purpose: Programming and triggering behavioral desire, subversion of psychic abilities of population, preparatory processing for mass electromagnetic control

Pseudonym: "Buzz Saw" E.E.M.C.

TOWER/ CIA, NSA:

Electronic cross country subliminal programming and suggestion

Targeting: Mass population, short-range intervals, long-range cumulative

Frequencies: Microwave, EHF SHF

Methodology: Cellular telephone system, ELF modulation

Purpose: Programming through neural resonance and encoded information

Effect: Neural degeneration, DNA resonance modification, psychic suppression

Pseudonym: "Wedding Bells"

B. More than 1500 6-month old Black and Hispanic babies in Los Angeles are given an "experimental" measles vaccine that had never been licensed for use in the United States. CDC later admits that parents were never informed that the vaccine being injected to their children was experimental.

1991

A. SRI's remote viewing project moved to SAIC

B. Desert Storm

According to Defense News, April 13-19, 1992, the US deployed an electromagnetic pulse weapon (EMP) in Desert Storm, designed to mimic the flash of electricity from a nuclear bomb. The Sandia National Laboratory had built a 23,000-square meter laboratory on the Kirkland Air Force Base in 1989 to house the Hermes II electron beam generator capable of producing 20 Trillion Watt pulses lasting 20-to-25 billionths-of-a-second. This X-ray simulator is called a "Particle Beam Fusion Accelerator". A stream of electrons hitting a metal plate can produce a pulsed X-ray or gamma ray. Hermes II had produced electron beams since 1974. These devices were apparently tested during the Gulf War, although detailed information on them is sparse.

1992

A. Eldon Byrd told me [Dick Farley] about it [lawsuit w/ Randi] over dinner at C. B. "Scott" Jones home one evening of several we spent together back in '92 and '93 there.

"Byrd said that Uri Geller put up $10,000 for his legal costs. Byrd and Geller are good friends, from back in the '70s..."

"Byrd says he had been "set-up" by postal inspectors - part of some initiative to discredit him because he was too public with his personal interests in "psi," etc. He'd allegedly had some Navy security clearance issues dog him, which contributed to his early retirement as one of their senior most civilian scientists." "When he was still with the Navy, Dr. Byrd was the contract manager for some of the research Michael Persinger did on 'neuro-impacts' of various EMFs and ELFs. Something about wave-propagation and influences on submariners if somebody "beeped" them with mind-influencing EMF signals, etc. That kind of thing." (Farley, Dick, "False Memory Spindrome")

B. John Alexander. Last year, Alexander organized a national conference devoted to researching "reports of ritual abuse, Near-Death Experiences, human contacts with extraterrestrial aliens, and other so-called anomalous experiences", the Albuquerque Journal reported in March 1993.

C. December.

"The U.S. Army's Armament Research, Development, and Engineering Center is conducting a one-year study of Acoustic Beam Technology ... The command awarded the 1-year study to Scientific Applications and Research Associates of Huntington Beach, CA. Related research is conducted at the Moscow-based Andreev Institute." (From "U.S. Explores Russian Mind Control Technology", by Barbara Opal, Defense News, Jan 11-17, 1993. Research by Harlan Girard and others.)

D. Dan Smith [the Aviary's physicist "Chicken Little"] was for a while engaged to marry the illustrious Rosemary Ellen Guiley, who numbered Von Ward, Jones, Colin Andrews, and a host of others on her "Center for Crop Circle Studies" advisory panel back in '92-93. Rosie's moved ahead (upward?) to the angelic realms, following the market and the money.

But she does have the corner on American expertise on Wicca, and her close relationship with former (now retired) Defense Intelligence Agency (DIA) "weird desk" and senior "applied anomalous" guy Dale Graff ... who was from my contacts with Scott Jones'. [Aviary's Prince Hans Adam- & Laurence Rockefeller-funded retired Naval Intel Cmdr. Cecil

B. "Scott" Jones] Affiliated with Sen. Claiborne Pell, with Dale Graff being the latter's "inside man" at the DIA. The "Aquarium Conspiracy" by Dan Smith and Rosemary Ellen Guiley

In the beginning, there was "eschatology" - the branch of theology dealing with the end times. Having spent many years first studying physics and then metaphysics, Dan came to the conclusion that the scientists have been looking at the World upside-down. Mind - not matter - is the foundation for all realities. Moreover, the materialist paradigm was in danger of imminent collapse, being subverted on the inside by its own contradictions and on the outside by the growing body of evidence for the paranormal. Creating and maintaining a reality is no easy game. It requires a lot of magic and a lot of conscious critters like ourselves who are pretty good at collective self-deception. Fortunately - or perhaps unfortunately - our particular reality game has about reached its natural conclusion, and we are waking up to the fact that mind and matter are not separate. We are undergoing an exciting-but-stressful revolution in our collective consciousness. This revolution - or global spiritual emergency - will bring upheavals and overloads in our global consciousness that will impact the material Earth for better-or-worse, and may quickly get out-of-control. We also will be opening up to other realities that will be impacting us as well. Our present very tidy sense of 'reality' and its boundaries is due to become much more fluid and permeable. Every spiritual tradition takes very seriously its prophesies about the End of the World. But for the first time, we are seeing these prophesies turning into believable predictions of Earth and reality changes. That is how Dan fell from physics into eschatology. [This last remark tells us that Rosemary is the one doing the writing here.] After experiencing numerous slammed doors among his former scientific colleagues, he decided that the most logical place to find fellow eschatologists would be in various intelligence agencies and among investigators of the paranormal. Dan next addressed how to communicate about the eschaton. Even a small hint that the

Government is worried about the End of the World might start a chain reaction of panic, which could possibly serve as a trigger for the eschaton.

On the other hand, people inside the Government might be wanting to set up a kind of civil defense network vis-a-vis the eschaton. And so, they would be looking for people on the outside who could much more freely network among the general public. An important link in the communication chain is what has become known as the "Aviary". This is the final link next to the public network, and so it must be heavily disguised by its own surrealistic smoke screen. The Aviary functions best by amplifying people's own misconceptions about the paranormal. It does this by helping to overinflate individual pieces of the puzzle so that particular investigators get pushed further into their own blind alleys. People are encouraged to be so "distracted by the trees that they fail to see the forest". This cacophony by people looking for truth in all the strangest places provides an excellent cover for the deadly serious business of clearing the decks and battening down the hatches for the eschaton event. It is like a Manhattan Project going on behind the scenes of alien grays and praying mantises having sex with humans. However, this eschaton conspiracy is being orchestrated by higher powers. And we don't mean the "Committee of 300". Very few of the people even near the center of the orchestration have a clear picture of what is coming down. But they do know that something is coming, and that they will have front-row seats. [note: Interesting that Guiley compares the government actions behind the scenes to the Manhattan Project, since that resulted in the most destructive weapon known to historical man up to that time. And now, subconsciously, Ms. Guiley has revealed to us the agenda: Total destruction of the masses of humanity except for those few elite who "help the project" and "sign agreements" to keep secrets - such as herself. "When you dance with the Devil, the Devil doesn't change - the Devil changes YOU.] The "Manhattan Project" relative to the eschaton is a global civilian network of people who will serve as a lightning rod for the cosmic energies coming in during the consciousness revolution. They will be looking to channel these energies into expanded realities. Thus, they will provide a degree of protection for those people who can find their places alongside the network.

Outside of the network, there will be greater levels of trauma and confusion. The pieces of the network are already in place. The remaining task is to properly activate and link the pieces into a critical mass of awareness. This last step is now underway. This is how an 'Aviary' helps to spawn an 'Aquarium', and how 'birds' learn to swim. The Aquarium is our business, and we are working to reach people who are ready to be activated in the consciousness revolution.

And here Ms. Guiley has described exactly what the Cassipeans have described for us and have helped us to uncover - that the Negative Hierarchy has created the New Age Movement, the Human Potential Movement, the "Contactee" and Gray Abduction Phenomenon for the express purpose of locking the planet into total Control - to make it a "headquarters" for the Lords of Darkness. And as the C's have said regarding the "Higher Powers" that Ms. Guiley reverentially refers to above:

A: At those levels, there is only one "Master."

Q: (L) "Those levels"? What 'levels'?

A: Levels that can hand down orders to bury or suppress. ... Those who are at that level have been bought-and-paid for by both giving knowledge of upcoming cataclysmic events, and promised survival and positions of power after. It is not difficult to realize that there is a body of such types in positions of power already. Power is not only attractive to such types, but they are also the kind most easily corrupted by it.

We also notice that Ms. Guiley has compared the process to that of "birds learning to swim". In other words, she and Mr. Smith are giving their signature as part of the Scale Gang – the Reptilian Overlords, as opposed to the Avian followers of the All Giving Mother.

According to Dan Smith - who undoubtedly is privy to a wealth of accurate though not widely known UFO information - this data is being released through him due to the grave concern by high government officials about impending metaphysical catastrophe –the eschaton, or the End of the World. What we see, however, is that Dan Smith and Rosemary Ellen Guiley are being used to further the agenda of the World Controllers who do, indeed, suspect the "End of the World", but have their own plans to survive it at the expense of most of the human population. It is this agenda that Ms. Guiley is now actively promoting in her writings and columns in Fate Magazine, which has become an organ of COINTELPRO, etc. We do notice that Ms. Guiley - like Whitley Strieber and the rest of the gang who play ball with the Matrix Controllers - have NO TROUBLE whatsoever getting "30 books published" and "hobnobbing" with the CIA.

1993

A. John Mack:

According to Dick Farley - former aide to C.B. Scott Jones - Laurance Rockefeller funneled "$194,000 to Mack's Harvard-affiliated Center for Psychology and Social Change via the Washington, D.C. chartered Human Potential Foundation, Inc. in the 1993-1994 period. Mack's group then started "PEER" (Program for Exceptional Experience Research) and operated an "alien abductee support group" who - among other functions they served - became fodder for Dr. Mack's 1994 Abductions. According to Donna Bassett - who infiltrated Mack's abductee support group - the Center for Psychology and Social Change (co-founded by Robert Jay Lifton) receives $250,000 a year from Rockefeller. Rockefeller also gave $194,000 to PEER along with various other donations. According to Bassett, Mack claims to have received funding from an ex-CIA source.

B. Kit Green.

C. Director of General Motor's Biomedical Research department. Attended closed meetings with Dr. Igor Smirnov under the auspices of his membership in the National Academy of Sciences panel on 21st Century Army Technologies. (Defense Electronics, 7/93. Reprinted in Flatland). Smirnov gave a series of closed meetings in Northern Virginia - starting on 3/17/93 - to the FBI, CIA, DIA, and ARPA concerning Russian developments with a device that allegedly implanted thoughts in a subject's mind. The FBI was considering using this device to implant the voice of God in David Koresh's mind, telling him to surrender. Other, non-intelligence participants included Dr. Richard Nakamura of the National Institute of Health [note: I think he may now be the director]. This technology was supposedly used by the Russians against civilians in Afghanistan and possibly on the Red Army to prepare them for battle. The American rights to this technology is owned by a Richmond, Virginia company called Psycotechnologies Corp. (Defense Electronics, 7/93. Reprinted in Flatland #11)

D. February 28, beginning of 51-day siege on the Branch Davidians at Waco Texas which ended in the death of more than 80 people. Until this incident, the electromagnetic weapons had kept a very low profile. But in the documentary video "Waco: The Big Lie Continues", footage from the British Broadcasting (BBC) shows at least 3 EM weapons used by U.S. Government agents. First, the noise generators used against the Davidians. Second, a powerful strobe light, shown during a nighttime sequence. And the third was the Russian psycho-acoustic weapon considered, but agents deny use of this weapon against the Waco people. FBI agents met with Dr. Igor Smirnov in Arlington, VA to discuss the possibility of using the weapon against the Davidians. (from "A Subliminal Dr. Strangelove", by Dorinda Elliot and John Barry, Newsweek, Aug 22, 1994). Janet Reno is also publicly connected to John Alexander (e.g., the recent "Non-Lethal Warfare" conference). See also: The Man Who Knew Too Much - What really happened at Waco? Carlos Ghigliotti thought he had the answer, and now he's dead. Was he a victim of conspiracy, or his own obsession? Ghigliotti - an expert in thermal imaging - was retained by the House Government Reform Committee last year to probe allegations that FBI agents - despite their vehement assertions to the contrary - had fired their weapons at members of the Branch Davidian sect, trapping helpless women and children inside the burning compound on April 19, 1993. Last fall, I had quoted him in The Post as saying that infrared surveillance tapes - as well as regular videos made by the Media - contained proof that the FBI fired. "The gunfire ... is there, without a doubt." In March, he was finalizing his report to Congress. He also had been advising attorneys waging a $100 million wrongful death suit against the Government on behalf of the Davidians and their heirs. "I still have a lot of shocking evidence to show you," he wrote in a March 28 letter to Michael Caddell, the lead attorney in that case. When his body was discovered, Ghigliotti's office got the scrutiny that Vince Foster's warranted after his suicide. Police sealed the premises and carted off computers and files. Rep. Dan Burton (R-Ind.) - whose committee had retained Ghigliotti - called for "a full and thorough investigation." The Justice Department's special counsel on Waco - John C. Danforth - asked a federal court to take control of all evidence from Ghigliotti's firm. I'd spent hours in that workshop reviewing tapes on his 8-monitor JVC video console, looking for evidence of Government perfidy in grainy images, debating theories while his beloved cats - Simone and Sipowicz - lolled at our feet. Carlos could be exasperating - brusque, inflexible and short-tempered, a fireplug of a guy who carried himself like a street fighter - but he had a soft side. More than once, he admitted to breaking down in tears while examining Waco evidence. Someone had to speak for the dead, he told me that he believed with all his heart that he had finally uncovered the Truth.

"I've solved the case," he announced during one of his calls in March, urging me to come once again to his lab to review videotapes. "I know exactly what happened."

But I was busy on other stories and never made it back. Now there was one more mystery to unravel: Was Carlos the final victim of Waco?

The Russian government is offering to share with the United States - in a bilateral Center for Psychotechnologies - the Soviet mind-control technology developed during the 1970s. The work was funded by the Department of Psycho-Correction at the Moscow Medical Academy.

"Acoustic psycho-correction involves the transmission of specific commands via static or white noise bands into the human subconscious ..." The Russian experts - among them former KGB General George Kotov - present in a paper a list of software and hardware available for $80,000. (From Opal article, "U.S. Explores Russian Mind Control Technology".)

E. High Frequency Active Auroral Research Program, HAARP
The HAARP is jointly managed by the US Air Force and the US Navy and is based in Gakona, Alaska. It is designed to "understand, simulate, and control ionospheric processes that might alter the performance of communication and surveillance systems". The HAARP system intends to beam 3.6 Gigawatts of effective radiated power of high- frequency radio energy into the ionosphere in order to (warning: all of which follows may be disinformation):

- Generate extremely low frequency (ELF) waves for communicating with submerged submarines

- Conduct geophysical probes to identify and characterize natural ionospheric processes so that techniques can be developed to mitigate or control them

- Generate ionospheric lenses to focus large amounts of high frequency energy, thus providing a means of triggering ionospheric processes that potentially could be exploited for Department of Defense purposes

- Electron acceleration for infrared (IR) and other optical emissions which could be used to control radio wave propagation properties

- Generate geomagnetic field aligned ionization to control the reflection/scattering properties of radio waves

- Use oblique heating to produce effects on radio wave propagation, thus broadening the potential military applications of ionospheric enhancement technology

F. During 1993, John Alexander and his team were working with Dr. Igor Smirnov

1994

A. By controlling the nature of the magnetic fields and causing them to simulate brain patterns, Persinger is able to stimulate strong emotions and hallucinations including the illusion of touch and movement. (Blackmore, Susan, "Alien

Abduction: The Inside Story", New Scientist, 11/19/94, pg. 29-31.) Persinger is shown demonstrating this device on the British TV show Horizon entitled "Close Encounters", written-and-narrated by Susan Blackmore. For their efforts, both Persinger and Blackmore have been accused of being in the Aviary.) Persinger was on the board of advisors for the False Memory Syndrome Foundation. Also, an informal advisor to SRI's remote-viewing program.

B. As of 1994, Smirnov has worked at Moscow's Institute of Psycho-Correction using subliminal technology as therapy for drug abusers and others. The Institute has been strapped for cash after the fall of the Soviet Union, but it has refused to accept business from the Russian Mafia. (Elliott, Dorinda and Barry, John, "A Subliminal Dr. Strangelove", Newsweek, 8/22/94, pg. 57) However, Smirnov works with the Human Potential Foundation and John Alexander. [note: I'm not sure how Smirnov's device is supposed to work.

Later reports claim it would work using inaudible, subliminal suggestions (spliced into phone conversations in the case of David Koresh). The device is definitely supposed to make the subject "hear" voices, as the FBI wanted to use Charlton Heston as the voice of God. This is definitely a different strategy from other subliminal techniques which are designed to produce mere suggestions.]

C. With a technique called "gene tracking," Dr. Garth Nicolson at the MD Anderson Cancer Center in Houston, TX discovers that many returning Desert Storm veterans are infected with an altered strain of Mycoplasma incognitos - a microbe commonly used in the production of biological weapons. Incorporated into its molecular structure is 40 percent of the HIV protein coat, indicating that it had been man-made.

D. Senator John D. Rockefeller (D-WV) issues a report revealing that for at least 50 years, the Department of Defense has used hundreds of thousands of military personnel in human experiments and for intentional exposure to dangerous substances. Materials included mustard and nerve gas, ionizing radiation, psychochemicals, hallucinogens, and drugs used during the Gulf War.

1995

A. James Randi:

"I told an audience at the annual meeting of the American Physical Society about the hilarious claims that Eldon Byrd made in court concerning important projects he'd been working on as a parapsychologist. One was a wristwatch that would protect the lucky wearer against the deadly effects of hairdryers and electric razors that bombard the brain with those 60-Hertz electrical waves. The watch would sense the phase of the offending waves and generate an opposing field to protect the subject.... But by far the best laugh of the trial was generated by Byrd when he proudly announced that as a result of reading and believing the book The Secret Life of

Plants he had a project going to train seaweed so that it could warn naval divers of danger." (Randi Hotline, 3/27/95)

B. SAIC - military contractor, located in California. [I couldn't find anything on remote viewing.] SAIC took over the research aspect of the remote-viewing program from SRI when director Ed May and his Cognitive Sciences Laboratory moved there in 1991. "SAIC - previously indicted on 10 felony fraud counts by the Justice Department relating to its management of a Superfund toxic cleanup site - has several prominent board members. Admiral Bobby Inman, former NSA Director and Deputy Director of the CIA; Melvin Laird, Richard Nixon's Defense Secretary; and retired General Max Thurman, Commander of the Panama Invasion. Previous board members include Robert Gates, former CIA Director; William Perry, current [1995] Secretary of Defense; and John Deutch, current [1995] CIA Director." SAIC owns Network Solutions, Inc. (NSI), which in September, 1995 took over control of Internet Domain Name registration from the National Science Foundation ("Spooks Spook Net Users", Paranoia, Issue 12, pg. 26). SAIC is also working with non-lethal weapons, but I haven't heard any details. (Brandt, Daniel, Mind Control and the Secret State). U.S. Government admits that it had offered Japanese war criminals and scientists - who had performed human medical experiments - salaries and immunity from prosecution in exchange for data on biological warfare research. (C) Dr. Garth Nicolson uncovers evidence that the biological agents used during the Gulf War had been manufactured in Houston, TX and Boca Raton, FL and tested on prisoners in the Texas Department of Corrections.

1996

A. NIDS established.

At one point, millionaire Robert Bigelow offered to provide funding to the tune of a million dollars for a cooperative research effort of the "Big Three" of ufology - MUFON

(Mutual UFO Network), CUFOS (Center for UFO Studies), and FUFOR (Fund for UFO Research). This effort - sometimes referred to as "the Coalition" - fell apart reportedly when Bigelow tried to control the direction of the group. UFO skeptic Philip Klass reportedly accused John Alexander of causing the break-up, although Alexander denies it. Maccabee has recently worked with the National Institute for Discovery Sciences (NIDS) and probably worked with-or-near Eldon Byrd, as the two both worked at the Naval Surface Weapons Center at the same time.

B. Courtney Brown, author of Cosmic Voyage, Dutton, 1996. Brown was trained in remote-viewing by Ed Dames and took courses at the Monroe Institute. Brown's book details his psychic conversations with aliens and repeats allegations similar to those made by Dames, Ingo Swann, Joe McMoneagle, and others. Among them are:

- Martians live among us and seek our help to return home. They live in South America and under the mountain Santa Fe Baldy.

- Brown psychically contacted Jesus, Guru Dev, and Buddha.

- The idea for the "Star Trek" television series was inspired by aliens to get humanity accustomed to the idea of working with alien races in a Federation.

- Specific plot elements of the "Star Trek: Next Generation" series were suggested to someone on the show via an implanted telepathy device.

Many of Dames' claims concerning the Martians are presented in Brown's book. But Brown implies that this is the first time any of this has been revealed to the remote-viewers, even though the sessions took place in 1994. Dames made similar claims as early as 1993 (see Stark, Debby, "Talking to Ed Dames", NM MUFON News, June/July 1993) Courtney Brown founded the Farsight in 1995. The Institute teaches a Scientific Remote Viewing course called "Farsight Voyager" which costs around $3,000.00. Here's the Institute's home page.

C. Radin is currently working with Joe McMoneagle in a project involved with remote-viewing future technology. (Compuserve On Line Conversation w/ Joe McMoneagle, 1/4/96)

D. Edgar Mitchell briefed then CIA director George Bush on the activities and results of the Instutute of Noetic Sciences (Mitchell, Edgar, The Way of the Explorer, GP Putnam's Sons, 1996, pg. 91).

E. Department of Defense admits that Desert Storm soldiers were exposed to chemical agents.

1997

A. In April, 1997, millionaire Robert Bigelow donated $3.7 million dollars to the University of Nevada to found the Bigelow Chair of Consciousness Studies, which allows students to take undergraduate courses dealing with parapsychology for college credit. These courses are related - though not formally linked - to Dean Radin's research at the University's Consciousness Research Laboratory which Bigelow once funded. Tart is currently teaching at the University of Las Vegas as part of Robert Bigelow's Bigelow Chair of Consciousness Studies (Patton, Natalie; "UNLV recruits authority in ESP", Las Vegas Review-Journal, 7/10/97).

B. 88 members of Congress sign a letter demanding an investigation into bioweapons use and Gulf War Syndrome.

1999

A. Michael Persinger:

"My research has not been 'funded by U.S. interests'. All of the money for our human research for the last 30 years has been from my personal income as a professor. The only funding ($10,000) we ever obtained from the U.S. was from the U.S. Navy - thanks to Eldon Byrd - to evaluate the effects of 0.5-Hz rotating magnetic fields upon the degranulation of mast cells in the rat brain. The effect was small but statistically significant." (Letter to Wes Thomas, 1/6/99)

B. 2000 Terminal experiments are being carried out on women, babies, and men of all ages.

SOURCES

Doc Hambone Good source, impeccably documented. Unfortunately relies a lot on McRae's Mind Wars which I have since found to be a little dubious. However, Hambone acknowledges this.

A background to the HAARP project. Prepared by Rosalie Bertell, Ph.D., GNSH

Health Freedom, Codex Issues Subject: Major Electromagnetic Mind Control Projects: Timeline to Present From: John Hammell

JH: A little flaky but interesting

Timeline: Electromagnetic Weapons by Judy Wall, Editor, Resonance Newsletter

Behind Closed Doors

Conversation w/ Joe McMoneagle, 1/4/96)

C. Edgar Mitchell briefed then CIA director George Bush on the activities and results of the Institute of Noetic Sciences (Mitchell, Edgar, The Way of the Explorer, GP Putnam's Sons, 1996, pg. 91).

D. Department of Defense admits that Desert Storm soldiers were exposed to chemical agents.

1997

A. In April, 1997, millionaire Robert Bigelow donated $3.7 million dollars to the University of Nevada to found the Bigelow Chair of Consciousness Studies, which allows students to take undergraduate courses dealing with parapsychology for college credit. These courses are related - though not formally linked - to Dean Radin's research at the University's Consciousness Research Laboratory which Bigelow once funded. Tart is currently teaching at the University of Las Vegas as part of Robert Bigelow's Bigelow Chair of Consciousness Studies (Patton, Natalie; "UNLV recruits authority in ESP", Las Vegas Review-Journal, 7/10/97).

B. 88 members of Congress sign a letter demanding an investigation into bioweapons use and Gulf War Syndrome.

1999

A. Michael Persinger:

"My research has not been 'funded by U.S. interests'. All of the money for our human research for the last 30 years has been from my personal income as a professor. The only funding ($10,000) we ever obtained from the U.S. was from the U.S. Navy - thanks to Eldon Byrd - to evaluate the effects of 0.5-Hz rotating magnetic fields upon the degranulation of mast cells in the rat brain. The effect was small but statistically significant." (Letter to Wes Thomas, 1/6/99)

B. 2000 Terminal experiments are being carried out on women, babies, and men of all ages.

SOURCES

Doc Hambone   Good source, impeccably documented. Unfortunately relies a lot on McRae's Mind Wars which I have since found to be a little dubious. However, Hambone acknowledges this.

A background to the HAARP project. Prepared by Rosalie Bertell, Ph.D., GNSH

Health Freedom, Codex Issues Subject: Major Electromagnetic Mind Control Projects: Timeline to Present From: John Hammell

JH: A little flaky but interesting.

Timeline: Electromagnetic Weapons by Judy Wall, Editor, Resonance Newsletter

Behind Closed Doors

It is virtually impossible to name everyone involved, or every agency's role, or country's specific role in an effort in such huge global magnitude. As a result, I am attempting to generalize and asked you Dear Reader, should you so desire, please fill in any blanks or gaps for yourself.

Dr. Robert Beck, an electronics engineer, is one former member of an elite corps of scientists recruited by his government to work in the area of electromagnetic frequencies. His published work talks of "specific frequencies that cause anxiety, fear and confusion." His unpublished work included "bizarre experiments in which would disorientate other scientists and is said to include changing moods from elation to depression. He described an experiment in which electromagnetic pulses emitted from a device that looked like a wristwatch caused other diners in a restaurant to talk more loudly or quietly, depending on the electromagnetic pulse emitted by the wristwatch device."

In 1963, Dr. Robert Becker explored effects of external magnetic-fields on brainwaves, showing a relationship between psychiatric hospital admissions and solar magnetic storms. He exposed volunteers to pulsed magnetic-fields similar to magnetic storms, and found a similar response. In the United States, sixty (60) Hz electric-power Extremely Low Frequencies (ELF) waves vibrate at the same frequency as the human brain. In the United Kingdom, fifty (50) Hz electricity emissions depress the thyroid.

Dr. R.O. Becker, twice nominated for the Nobel Prize for his health work in bio-electromagnetism was more explicit in his concern over illicit government activity later. He wrote of "obvious applications in covert operations designed to drive a target crazy with "voices." The 1976 DIA report also credits the Soviets with other capabilities, stating, "Sounds and possibly even words which appear to be originating intercranially can be induced by signal modulations at very low power densities." The Body Electric- Electromagnetism and the Foundation of Life, by Robert O. Becker, M.D. and Gary Selden, (Quill Publisher).

Andrija Puharich, MD, original name is Henry K. Puharich, (February 19, 1918 - January 3, 1995) was an Army officer in the early 1950s. During that time, made frequent visits Edgewood Arsenal Research Laboratories and Camp Detrick, meeting with various high-ranking officers and officials, primarily from the Pentagon, CIA, and Naval Intelligence. The Edgewood Arsenal is currently officially called the Edgewood Area of Aberdeen Proving Ground. Puharich was a medical and parapsychological researcher, medical inventor and author, who is, perhaps best known as the person who brought Israeli Uri Geller and Peter Hurkos, to the United States for parapsychology investigation. Uri Geller was an Israeli born, self-proclaimed psychic living in England known for his trademark television performances of spoon bending and other supposed psychic effects. Peter Hurkos, was a Dutch psychic and he also investigated Mexican psychic surgeon Pachita.

Dr. Andrija Puharich (in the 1950 & 60s), found that a clairvoyant's brainwaves turned to 8 Hz when their psychic powers were operative. In 1956, he observed an Indian Yogi controlling his brainwaves, deliberately shifting his consciousness from one level to another. Puharich trained people via bio-feedback to do this consciously, that is, creating 8 Hz waves with the technique of bio-feedback. A psychic healer generated 8 Hz waves through a hands-on healing process, actually alleviating that patient's heart trouble; the healer's brain emitting 8 Hz.

One person, emitting a certain frequency, can make another also resonate to the same frequency. Our brains are extremely vulnerable to any technology that sends out ELF waves, because they immediately start resonating to the outside signal by a kind of tuning-fork effect.

Puharich further experimented, discovering that, 7.83 Hz (earth's pulse rate) made a person "feel good," producing an altered-state 10.80 Hz causes riotous behavior 6.6 Hz causes depression.

Puharich made ELF waves change RNA and DNA in the body, breaking hydrogen bonds to make a person resonate at a higher vibratory rate. He really wanted to go beyond the psychic 8 Hz brainwave and attract psi phenomena.

James Hurtak, who once worked for Puharich, also wrote in his book, "The Keys of Enoch" that ultra-violet caused hydrogen bonds to break and this raised the vibratory rate.

Puharich presented the mental effects of ELF waves to military leaders, but they would not believe him. He then gave this information to certain dignitaries of other Western nations. The United States Government burned down his home in New York to shut him up, whereas he then fled to Mexico. However, the Russians discovered which ELF frequencies affected a certain portion of the human brain. It was on July 4, 1976, that they began zapping the U.S. Embassy in Moscow with electromagnetic-waves, varying the signal, also focusing on 10 Hz. (10 Hz puts people into a hypnotic state). Russians and North Koreans use this in portable mind-control machines to extract confessions. (This system can also be found in some American Churches!)

This Russian "Woodpecker" signal was traveling across the world from a transmitter near Kiev. The U.S. Air Force identified 5 different frequencies in this compound that the harmonic Woodpecker was sending through the earth and atmosphere.

In Mexico, Puharich continued to monitor the Russian ELF wave signal and the higher harmonics (5.340 MHz) in the MHz range. He was somehow induced to work for the CIA and he and Dr. Robert Becker designed equipment to measure these waves and their effect on the human brain. Puharich started his work by putting dogs to sleep.

By 1948/49, he had graduated to monkeys, deliberately destroying their eardrums to enable them to pick up sounds without the eardrum intact. He discovered a nerve from the tongue could be used to facilitate hearing. He created the tooth implant that mind-control victims are now claiming was put in by their dentist, unbeknownst to them, and causing them to hear "voices in their head.' These were placed under caps or lodged in the jaw as the technology continued it agenda by various nations for political control of the human mind.

CHAPTER TEN

"The enemy of the moment always represented absolute evil, and it followed that any past or future agreement with him was impossible."

- George Orwell (1984)

The plans to create a mind controlled workers society ideation has been in place for a long time. The desire and current technology grew out of experiments that the Nazis started before World War II and intensified during the time of the Nazi concentration camps

As discussed earlier, the Soviets had long understood the powerful biological impacts of radio frequency (RF) and microwave (MW) technologies. In the early 1950's they were becoming increasingly aware of the non-thermal bio effects from microwaves from years of research prior to. Historically the Soviet Union has invested huge sums of money and time investigating microwave remote brain manipulation and their effects and programs in energetics and psycho-energetics technology, known to the West as psychotronics. The bulk of the initial work on the science underpinning this technology had been done in the West and smuggled to the Soviet Union. This included some of Tesla findings which also evolved in the Hitler regime's research programs. For decades, the scientific community of the West had ignored the work of people like Moray Abrams, Hieronymous, Tesla, De la Warr, Down and Reich, giving the Soviets at least a 30-year head start to build to consolidate their position in psychotronic weaponry. When Brezhnev suggested at the 1978 SALT negotiations banning weapons "more frightful than the mind of man has ever conceived," President Carter had no idea what he really was suggesting.

Dwight D. Eisenhower's presidency (1953 to 1960) was coming to an end in late 1960. In his last speech, he puzzled some by appearing to be issuing some sort of coded warning. And the undercurrent, though not specifically stated appeared to be directly related to new technology called the "Sound of Silence or Sound of Silence Spread Spectrum (SSSS.) Today, research has proven that when it is coupled with the new all-digital TV signal that it could very well be a medium of delivery of various subliminal messages into the minds of an unsuspecting U.S. populace. The fact is it can be deployed through HAARP and GWEN Towers.

Eisenhower was an honest and dedicated patriotic American. Like Marine Corp General Smedley Butler, who decades earlier declared to Congress that "War is a Racket" --- Ike knew that such absolute power and total covert control over the minds and hearts of individual citizens would corrupt society absolutely. He also knew and understood, as did the German philosopher Goethe:

"No man is more hopelessly enslaved, than he who falsely believes that he is free."

Therefore, he issued his strong, concluding warning to America. The Sound of Silence is a military-intelligence code word for certain psychotronic weapons of mass mind-control tested in the mid 1950's, perfected during the 70's, and used extensively by the "modern" U.S. military in the early 90's, despite the opposition and warnings issued by men such as Dwight David Eisenhower.

This mind-altering covert weapon is based on something called subliminal carrier technology, or the Silent Sound Spread Spectrum (SSSS), (also nicknamed S-Quad or "Squad" in military jargon.) It was developed for military use by Dr. Oliver Lowery of Norcross, Georgia, and is described in US Patent #5,159,703 – "Silent Subliminal Presentation System" for commercial use in 1992.

The patent abstract reads:

"A silent communications system in which non-aural carriers, in the very low (ELF) or very high audio-frequency (VHF) range or in the adjacent ultrasonic frequency spectrum are amplitude –or frequency-modulated with the desired intelligence and propagated acoustically vibrationally, for inducement in to the brain, typically through the use of loudspeakers, earphones, or piezoelectric transducers. The modulated carriers may be transmitted directly in real-time or may be conveniently recorded and stored on mechanical, magnetic, or optical media for delayed or repeated transmission to the listener."

In layman's terms, this device, this "Sound of Silence" simply allows for the unwarranted implantation of specific thoughts, emotions, and even prescribed physical actions into unsuspecting human beings.

In short, it has the very real ability to turn human beings into mere puppets in the hands of certain "controllers" or puppet-masters. Eisenhower knew full well what such a "weapon" could do in the hands of greedy, conspiring men and women scheming to control the planet. It could easily result in the takeover of American society without a single bullet being fired

Above excerpts from the Sound of Silence -The Antithesis of Freedom, By A. True Ott, PhD, ND

Dwight D. Eisenhower

Farewell Address EXCERPT
Delivered 17 January 1961

"In the councils of government, we must guard against the acquisition of unwarranted influence, whether sought or unsought, by the military-industrial complex. The potential for the disastrous rise of misplaced power exists and will persist. We must never let the weight of this combination endanger our liberties or democratic processes.

We should take nothing for granted. Only an alert and knowledgeable citizenry can compel the proper meshing of the huge industrial and military machinery of defense with our peaceful methods and goals, so that security and liberty may prosper together..."

Eisenhower's predecessor, John F. Kennedy, once taking office, almost expressed similar concerns, some would argue, surrounding some type of technological agenda.

John F. Kennedy, Jr. speech, April 27, 1961, on Secret Societies said:

"We are opposed around the world by a monolithic and ruthless conspiracy that relies primarily on covert means for expanding its sphere of influence--on infiltration instead of invasion, on subversion instead of elections, on intimidation instead of free choice, on guerrillas by night instead of armies by day. It is a system which has conscripted vast human and material resources into the building of a tightly knit, highly efficient machine that combines military, diplomatic, intelligence, economic, scientific and political operations."

All across the military, there is interest in translating thoughts into computer code, and vice-versa. Today, DARPA funded researchers have taught monkeys how to control robotic limbs with their thoughts. Defense contractor, such as Northrup Grumman, is building binoculars that tap the unconscious mind. Honeywell has built a system that monitors pre-conscious neural firings, to help pick out targets in satellite imagery. The JASON's, the Pentagon's premiere scientific advisory board, has warned of the dangers of enemies implanted with brain-computer interfaces. And the Defense Intelligence Agency just released a report, saying the military needs to spend more on neuroscience – up to and including "making the enemy obey our commands. The problem is the line between who the enemy really is has become blurred and technology is now turned on its and operated by those practicing unscrupulous, unethical tactics and strategies due to little public knowledge and as I write in "You Are Not My "Big Brother!" the covert, subliminal, manipulative capabilities of numerous technologies and numerous delivery systems. In other words, "All Hell has broken loose" as the ethics and moral issues surrounding the technology and inevitable abuses and victimizations, outside of any laws of protection, continue to escalate unchecked and unmonitored.

In the CBS 60 Minutes documentary on mind reading technology, June 28, 2009, the reality of testing, development and use of this specific technology is understated. The ideation for mind reading, is a fact today by technology effectively delivered by satellite radar, optical eye / laser beam.

Research, testing and studies continued around to the Electroencephalograph (EEG) invention of 1920. Many years later, Lawrence Pinneo, a neurophysiologist and electronic engineer working for Stanford Research Institute (a military contractor as stated earlier) became known as the first pioneer in this field. In 1974, he developed a computer system which correlated brain waves on an electroencephalograph with specific commands which was patented as the Brain Wave Monitor / Analyzer. Science now had the ability to decipher human thought, reading the human mind and later from a great distance as seen below through Brainwave Scanner / Programs:

Later, in the early 1990s, Dr. Edward Taub reported that words could be communicated onto a screen using the thought-activated movements of the computer cursor. The first program developed in 1994 by Dr. Donald York and Dr. Thomas Jensen.

The brain wave patterns of 40 subjects were officially correlated with both spoken words and silent thought. This was achieved by a neurophysiologist, Dr. Donald York, and a speech pathologist, Dr. Thomas Jensen, from the University of Missouri. They then clearly identified 27 words and syllables in specific brain wave patterns and produced a computer program with a brain wave vocabulary. Today, using satellite lasers, and high-powered computers, the agencies gained the ability to decipher human thoughts - from a considerable distance (instantaneously.) The computer program had a vocabulary of over 60,000 words and phrases in many different languages at one time, which logically, has continued it accumulation.

The brain-computer interface would use a noninvasive brain imaging technology like electroencephalography (EEG) to let people communicate thoughts to each other. For example, a soldier would "think" a message to be transmitted and a computer-based speech recognition system would decode the EEG signals. The decoded thoughts, in essence translated brain waves, are transmitted using a system that points in the direction of the intended target.

Practically, EEG technology could be used to communication with stroke victims and brain-activated control of modern jets. These are two applications. However, more often, it is used to mentally rape a Civilian target; their thoughts being referenced immediately and/or recorded for future use.

HETERODYNE EEG CLONING

DESCRIPTION: A system whereby the target's EMF is monitored remotely and EEG results fed back to them (or others) to mimic emotional patterns; e.g. fear, anger, etc..

PURPOSE: To induce emotional / psychological responses. For example, the feedback of Delta waves may induct drowsiness since these are familiar when in deep sleep.

NOTE: In conjunction with Neurophone technology, the Malech technology, below, is a powerful, subliminal mechanism for remote interrogations / torture via satellite.

MALECH'S REMOTE BRAINWAVE ALTERING MACHINE

US Patent # 3951134 (April 1976) by Robert G. Malech

Apparatus for and method of sensing brain waves at a position remote from a subject whereby electromagnetic signals of different frequencies are simultaneously transmitted to the brain of the subject in which the signals interfere with one another to yield a waveform which is modulated by the subject's brain waves. The interference waveform which is representative of the brain wave activity is re-transmitted by the brain to a receiver where it is demodulated and amplified. The demodulated waveform is then displayed for visual viewing and routed to a computer for further processing and analysis. The demodulated waveform also can be used to produce a compensating signal which is transmitted back to the brain to effect a desired change in electrical activity therein.

This device can not only read brainwaves, but also alter them from a distance!

Adding a computer trained to associate various sensory experiences with corresponding patterns of brainwave activity could produce a device capable of making people hear voices, smell odors, see images, feel sensations, and even experience emotions imposed by someone else. Mind reading technology is now capable of deciphering human, or other patented abilities thought to the letter.

Below is a small list of other United States Patent and Trademark Office official patents for psychological electronic technology:

US Patent # 5,017,143 (May 21, 1991) Method and Apparatus for Producing Subliminal Images

Backus, Alan, et al. Abstract - A method and apparatus to produce more effective visual subliminal communications. Graphic and/or text images, presented for durations of less than a video frame, at organized rhythmic intervals, the rhythmic intervals intended to affect user receptivity, moods or behavior. Subliminal graphic images having translucent visual values locally dependent on background values in order to maintain desired levels of visual contrast.)

- US Patent - 5,629,678 Implantable Transceiver (Microchip)

- US Patent – 5878155 RFID Barcode Tattoo on humans

- US Patent - 5,539,705 Ultrasonic Speed Translator and Communications System

- US Patent - 5,629,678 Personal Tracking and Recovery System

- US Patent - 5,760,692 Intro - Oral Tracking Device

- US Patent - 5,270,800 Subliminal Message Generator

- US Patent - 4,877,027 Hearing System (Microwave Voice to Skull)

- US Patent - 3,837,331 (September 24, 1974) System & Method for Controlling the Nervous System of a Living Organism

- US Patent - 5,507,291 Method and an Associated Apparatus for Remotely Determining Information as to a Person's Emotional State

- US Patent - 4,777,529 (October 11, 1988) Auditory Subliminal Programming System

- US Patent - 4,717,343 (January 5, 1988) Method of Changing a Person's Behavior

METHOD AND SYSTEMS FOR ALTERING CONSCIOUSNESS:

1. U.S. Patent, #5,123,899. June 23, 1992. This is a system for stimulating the brain to exhibit specific brain wave rhythms and thereby altering the subjects' state of consciousness.

2. U.S. Patent, #5,289,438. February 22, 1994.

DESCRIPTION: A system for the simultaneous application of multiple stimuli (usually aural) with different frequencies and waveforms.

Electro Magnetic Field (EMF) monitoring / interference is one of the most insidious and secretive of all methods used by the agencies and defense contractors. Similarly, EEG cloning feeds back the results of EMF monitoring in an attempt to induce emotional responses (e.g. fear, anger, even sleep, etc.)

US Patent # 6,506,148 (January 14, 2003) Nervous System Manipulation by EM Fields from Monitors

Loos, Hendricus Abstract --- Physiological effects have been observed in a human subject in response to stimulation of the skin with weak electromagnetic fields that are pulsed with certain frequencies near 1/2 Hz or 2.4 Hz, such as to excite a sensory resonance. Many computer monitors and TV tubes, when displaying pulsed images, emit pulsed electromagnetic fields of sufficient amplitudes to cause such excitation. It is therefore possible to manipulate the nervous system of a subject by pulsing images displayed on a nearby computer monitor or TV set. For the latter, the image pulsing may be imbedded in the program material, or it may be overlaid by modulating a video stream, either as an RF signal or as a video signal. The image displayed on a computer monitor may be pulsed effectively by a simple computer program. For certain monitors, pulsed electromagnetic fields capable of exciting sensory resonances in nearby subjects may be generated even as the displayed images are pulsed with subliminal intensity.

US Patent # 6,488,617 (December 3, 2002) Method and Device for Producing a Desired Brain State

Katz, Bruce Abstract --- A method and device for the production of a desired brain state in an individual contain means for monitoring and analyzing the brain state while a set of one or more magnets produce fields that alter this state. A computational system alters various parameters of the magnetic fields in order to close the gap between the actual and desired brain state. This feedback process operates continuously until the gap is minimized and/or removed.

US Patent # 6,487,531 (November 26, 2002) Signal Injection Coupling into the Human Vocal Tract

Tosaya, Carol Abstract --- A means and method are provided for enhancing or replacing the natural excitation of the human vocal tract by artificial excitation means, wherein the artificially created acoustics present additional spectral, temporal, or phase data useful for (1) enhancing the machine recognition robustness of audible speech or (2) enabling more robust machine-recognition of relatively inaudible mouthed or whispered speech. The artificial excitation (a) may be arranged to be audible or inaudible, (b) may be designed to be non-interfering with another user's similar means, (c) may be used in one or both of a vocal content-enhancement mode or a complimentary vocal tract-probing mode, and/or (d) may be used for the recognition of audible or inaudible continuous speech or isolated spoken commands.

US Patent # 6,426,919 (July 30, 2002) Portable and Hand-Held Device for Making Humanly Audible Sounds

Gerosa, William Abstract --- A portable and hand-held device for making humanly audible sounds responsive to the detecting of ultrasonic sounds. The device includes a hand-held housing and circuitry that is contained in the housing. The circuitry includes a microphone that receives the ultrasonic sound, a first low voltage audio power amplifier that strengthens the signal from the microphone, a second low voltage audio power amplifier that further strengthens the signal from the first low voltage audio power amplifier, a 7-stage ripple carry binary counter that lowers the frequency of the signal from the second low voltage audio power amplifier so as to be humanly audible, a third low voltage audio power amplifier that strengthens the signal from the 7-stage ripple carry binary counter, and a speaker that generates a humanly audible sound from the third low voltage audio power amplifier.

US Patent # 6,292,688 (September 18, 2001) Method and Apparatus for Analyzing Neurological Response to Emotion-Inducing Stimuli

Patton, Richard Abstract --- A method of determining the extent of the emotional response of a test subject to stimuli having a time-varying visual content, for example, an advertising presentation. The test subject is positioned to observe the presentation for a given duration, and a path of communication is established between the subject and a brain wave detector/analyzer. The intensity component of each of at least two different brain wave frequencies is measured during the exposure, and each frequency is associated with a particular emotion. While the subject views the presentation, periodic variations in the intensity component of the brain waves of each of the particular frequencies selected is measured. The change rates in the intensity at regular periods during the duration are also measured. The intensity change rates are then used to construct a graph of plural coordinate points, and these coordinate points graphically establish the composite emotional reaction of the subject as the presentation continues.

US Patent # 6,258,022 (July 10, 2001) Behavior Modification

Rose, John Abstract --- Behavior modification of a human subject takes place under hypnosis, when the subject is in a relaxed state. A machine plays back a video or audio recording, during which the subject is instructed to activate a device to create a perceptible stimulation which is linked, through the hypnosis, with a visualization of enhanced or improved performance. After the hypnosis, the user can reactivate the device at will, whenever the improved performance, such as an improved sporting performance, is desired. This will again create the perceptible stimulation and thus induce the required visualization.

US Patent # 6,239,705 (May 29, 2001) Intra-Oral Electronic Tracking Device

Glen, Jeffrey Abstract --- An improved stealthy, non-surgical, bio-compatible electronic tracking device is provided in which a housing is placed intra-orally. The housing contains micro-circuitry. The micro-circuitry comprises a receiver, a passive mode to active mode activator, a signal decoder for determining positional fix, a transmitter, an antenna, and a power supply. Optionally, an amplifier may be utilized to boost signal strength. The power supply energizes the receiver. Upon receiving a coded activating signal, the positional fix signal decoder is energized, determining a positional fix. The transmitter subsequently transmits through the antenna a position locating signal to be received by a remote locator. In another embodiment of the present invention, the micro-circuitry comprises a receiver, a passive mode to active mode activator, a transmitter, an antenna and a power supply. Optionally, an amplifier may be utilized to boost signal strength. The power supply energizes the receiver. Upon receiving a coded activating signal, the transmitter is energized. The transmitter subsequently transmits through the antenna a homing signal to be received by a remote locator.

US Patent # 6,167,304 (December 26, 2000) Pulse Variability in Electric Field Manipulation of Nervous Systems

Loos, Hendricus Abstract --- Apparatus and method for manipulating the nervous system of a subject by applying to the skin a pulsing external electric field which, although too weak to cause classical nerve stimulation, modulates the normal spontaneous spiking patterns of certain kinds of afferent nerves. For certain pulse frequencies, the electric field stimulation can excite in the nervous system resonances with observable physiological consequences. Pulse variability is introduced for the purpose of thwarting habituation of the nervous system to the repetitive stimulation, or to alleviate the need for precise tuning to a resonance frequency, or to control pathological oscillatory neural activities such as tremors or seizures. Pulse generators with stochastic and deterministic pulse variability are disclosed, and the output of an effective generator of the latter type is characterized.

US Patent # 6,135,944 (October 24, 2000) Method of Inducing Harmonious States of Being

Bowman, Gerard D., et al. Abstract --- A method of inducing harmonious states of being using vibrational stimuli, preferably sound, comprised of a multitude of frequencies expressing a specific pattern of relationship. Two base signals are modulated by a set of ratios to generate a plurality of harmonics. The harmonics are combined to form a "fractal" arrangement.

US Patent # 6,122,322 (September 19, 2000) Subliminal Message Protection

Jandel, Magnus Abstract --- The present invention relates to a method and to a system for detecting a first context change between two frames. When a second context change between a further two frames occurs within a predetermined time interval, the frames accommodated within the two context changes are defined as a subliminal message. An alarm is sent to an observer upon detection of a subliminal message.

US Patent # 6,091,994 (July 18, 2000) Pulsative Manipulation of Nervous Systems

Loos, Hendricus Abstract --- Method and apparatus for manipulating the nervous system by imparting subliminal pulsative cooling to the subject's skin at a frequency that is suitable for the excitation of a sensory resonance. At present, two major sensory resonances are known, with frequencies near 1/2 Hz and 2.4 Hz. The 1/2 Hz sensory resonance causes relaxation, sleepiness, ptosis of the eyelids, a tonic smile, a "knot" in the stomach, or sexual excitement, depending on the precise frequency used. The 2.4 Hz resonance causes the slowing of certain cortical activities, and is characterized by a large increase of the time needed to silently count backward from 100 to 60, with the eyes closed. The invention can be used by the general public for inducing relaxation, sleep, or sexual excitement, and clinically for the control and perhaps a treatment of tremors, seizures, and autonomic system disorders such as panic attacks. Embodiments shown are a pulsed fan to impart subliminal cooling pulses to the subject's skin, and a silent device which induces periodically varying flow past the subject's skin, the flow being induced by pulsative rising warm air plumes that are caused by a thin resistive wire which is periodically heated by electric current pulses.

US Patent # 6,081,744 (June 27, 2000) Electric Fringe Field Generator for Manipulating Nervous Systems

Loos, Hendricus Abstract --- Apparatus and method for manipulating the nervous system of a subject through afferent nerves, modulated by externally applied weak fluctuating electric fields, tuned to certain frequencies such as to excite a resonance in neural circuits. Depending on the frequency chosen, excitation of such resonances causes in a human subject relaxation, sleepiness, sexual excitement, or the slowing of certain cortical processes. The electric field used for stimulation of the subject is induced by a pair of field electrodes charged to opposite polarity and placed such that the subject is entirely outside the space between the field electrodes. Such configuration allows for very compact devices where the field electrodes and a battery-powered voltage generator are contained in a small casing, such as a powder box. The stimulation by the weak external electric field relies on frequency modulation of spontaneous spiking patterns of afferent nerves. The method and apparatus can be used by the general public as an aid to relaxation, sleep, or arousal, and clinically for the control and perhaps the treatment of tremors and seizures, and disorders of the autonomic nervous system, such as panic attacks.

US Patent # 6,052,336 (April 18, 2000) Apparatus and Method of Broadcasting Audible Sound Using Ultrasonic Sound as a Carrier

Lowrey, Austin, III Abstract --- An ultrasonic sound source broadcasts an ultrasonic signal which is amplitude and/or frequency modulated with an information input signal originating from an information input source. If the signals are amplitude modulated, a square root function of the information input signal is produced prior to modulation. The modulated signal, which may be amplified, is then broadcast via a projector unit whereupon an individual or group of individuals located in the broadcast region detect the audible sound.

US Patent # 6,039,688 (March 21, 2000) Therapeutic Behavior Modification Program, Compliance Monitoring and Feedback System

Douglas, Peter, et al. Abstract --- A therapeutic behavior modification program, compliance monitoring and feedback system includes a server-based relational database and one or more microprocessors electronically coupled to the server. The system enables development of a therapeutic behavior modification program having a series of milestones for an individual to achieve lifestyle changes necessary to maintain his or her health or recover from ailments or medical procedures. The program may be modified by a physician or trained case advisor prior to implementation. The system monitors the individual's compliance with the program by prompting the individual to enter health-related data, correlating the individual's entered data with the milestones in the behavior modification program and generating compliance data indicative of the individual's progress toward achievement of the program milestones. The system also includes an integrated system of graphical system interfaces for motivating the individual to comply with the program. Through the interfaces, the individual can access the database to review the compliance data and obtain health information from a remote source such as selected sites on the Internet. The system also provides an electronic calendar integrated with the behavior modification program for signaling the individual to take action pursuant to the behavior modification program in which the calendar accesses the relational database and integrates requirements of the program with the individual's daily schedule, and an electronic journal for enabling the individual to enter personal health-related information into the system on a regular basis. In addition, the system includes an electronic meeting room for linking the individual to a plurality of other individuals having related behavior modification programs for facilitating group peer support sessions for compliance with the program. The system enables motivational media presentations to be made to the individuals in the electronic meeting room as part of the group support session to facilitate interactive group discussion about the presentations. The entire system is designed around a community of support motif including a graphical electronic navigator operable by the individual to control the microprocessor for accessing different parts of the system.

US Patent # 6,017,302 (January 25, 2000) Subliminal Acoustic Manipulation of Nervous Systems (sexual arousal)

Loos, Hendricus Abstract --- In human subjects, sensory resonances can be excited by subliminal atmospheric acoustic pulses that are tuned to the resonance frequency. The 1/2 Hz sensory resonance affects the autonomic nervous system and may cause relaxation, drowsiness, or sexual excitement, depending on the precise acoustic frequency near 1/2 Hz used. The effects of the 2.5 Hz resonance include slowing of certain cortical processes, sleepiness, and disorientation. For these effects to occur, the acoustic intensity must lie in a certain deeply subliminal range. Suitable apparatus consists of a portable battery-powered source of weak sub-audio acoustic radiation. The method and apparatus can be used by the general public as an aid to relaxation, sleep, or sexual arousal, and clinically for the control and perhaps treatment of insomnia, tremors, epileptic seizures, and anxiety disorders. There is further application as a nonlethal weapon that can be used in law enforcement standoff situations, for causing drowsiness and disorientation in targeted subjects. It is then preferable to use venting acoustic monopoles in the form of a device that inhales and exhales air with sub-audio frequency.

US Patent # 6,011,991 (January 4, 2000) Communication System & Method Including Brain Wave Analysis

Mardirossian, Aris Abstract --- A system and method for enabling human beings to communicate by way of their monitored brain activity. The brain activity of an individual is monitored and transmitted to a remote location (e.g. by satellite). At the remote location, the monitored brain activity is compared with pre-recorded normalized brain activity curves, waveforms, or patterns to determine if a match or substantial match is found. If such a match is found, then the computer at the remote location determines that the individual was attempting to communicate the word, phrase, or thought corresponding to the matched stored normalized signal.

US Patent # 6,006,188 (December 21, 1999) Speech Signal Processing for Determining Psychological or Physiological Characteristics

Bogdashevsky, Rostislav, et al. Abstract --- A speech-based system for assessing the psychological, physiological, or other characteristics of a test subject is described. The system includes a knowledge base that stores one or more speech models, where each speech model corresponds to a characteristic of a group of reference subjects. Signal processing circuitry, which may be implemented in hardware, software and/or firmware, compares the test speech parameters of a test subject with the speech models. In one embodiment, each speech model is represented by a statistical time-ordered series of frequency representations of the speech of the reference subjects. The speech model is independent of a priori knowledge of style parameters associated with the voice or speech. The system includes speech parameterization circuitry for generating the test parameters in response to the test subject's speech. This circuitry includes speech acquisition circuitry, which may be located remotely from the knowledge base. The system further includes output circuitry for outputting at least one indicator of a characteristic in response to the comparison performed by the signal processing circuitry. The characteristic may be time-varying, in which case the output circuitry outputs the characteristic in a time-varying manner. The output circuitry also may output a ranking of each output characteristic. In one embodiment, one or more characteristics may indicate the degree of sincerity of the test subject, where the degree of sincerity may vary with time. The system may also be employed to determine the effectiveness of treatment for a psychological or physiological disorder by comparing psychological or physiological characteristics, respectively, before and after treatment.

US Patent # 5,954,630 (September 21, 1999) FM Theta-Inducing Audible Sound

Masaki, Kazumi, et al. Abstract --- An audible sound of modulated wave where a very low-frequency wave of about 20 hertz or lower is superposed on an audio low-frequency wave effectively stimulates Fm theta in human brain waves to improve attention and concentration during mental tasks when auditorily administered. The audible sound is also effective in stimulation of human alpha wave when the very low-frequency wave lies within the range of about 2-10 hertz. Such audible sound is artificially obtainable by generating an electric signal which contains such a modulated wave, and transducing it into audible sound wave.

US Patent # 5,954,629 (September 21, 1999) Brain Wave Inducing System

Yanagidaira, Masatoshi, et al. Abstract --- Sensors are provided for detecting brain waves of a user, and a band-pass filter is provided for extracting a particular brain waves including an alpha wave included in a detected brain wave. The band-pass filter comprises a first band-pass filter having a narrow pass band, and a second band-pass filter having a wide pass band. One of the first and second band-pass filters is selected, and a stimulation signal is produced in dependency on an alpha wave extracted by a selected band-pass filter. In accordance with the stimulation signal, a stimulation light is emitted to the user in order to induce the user to relax or sleeping state.

US Patent # 5,935,054 (August 10, 1999) Magnetic Excitation of Sensory Resonances

Loos, H. Abstract --- The invention pertains to influencing the nervous system of a subject by a weak externally applied magnetic field with a frequency near 1/2 Hz. In a range of amplitudes, such fields can excite the 1/2 sensory resonance, which is the physiological effect involved in "rocking the baby".

US Patent # 5,922,016 (July 13, 1999) Apparatus for Electric Stimulation of Auditory Nerves of a Human Being

Wagner, Hermann Abstract --- Apparatus for electric stimulation and diagnostics of auditory nerves of a human being, e.g. for determination of sensation level (SL), most conformable level (MCL) and uncomfortable level (UCL) audibility curves, includes a stimulator detachably secured to a human being for sending a signal into a human ear, and an electrode placed within the human ear and electrically connected to the stimulator by an electric conductor for conducting the signals from the stimulator into the ear. A control unit is operatively connected to the stimulator for instructing the stimulator as to characteristics of the generated signals being transmitted to the ear.

US Patent # 5,868,103 (February 9, 1999) Method and Apparatus for Controlling an Animal

Boyd, Randal, Abstract --- An apparatus for controlling an animal wherein the animal receives a control stimulus of the release of a substance having an adverse effect upon the animal as a corrective measure. The apparatus includes a transmitter for producing a transmitted field, and a releasable collar for attaching to the neck of the animal. The collar includes a receiver for receiving the transmitted field and for producing a received signal, a control circuit for determining when the received signal indicates that the animal requires a corrective measure and for producing a control signal, a container for containing the substance having an adverse effect upon the animal, and a mechanism for releasing the substance from the container into the presence of the animal upon the production of the control signal by the control circuit. In use, the transmitter is set to produce the transmitted field and the collar is attached to the neck of the animal.

As the animal moves about, the receiver in the collar receives the transmitted field and produces a received signal. The control circuit determines when the received signal indicates that the animal requires a corrective measure. A control signal is produced by the control circuit when the determination is made that the animal requires a corrective measure. Upon the production of the control signal, the substance having an adverse effect upon the animal is released from the container and into the presence of the animal.

US Patent # 5,784,124 (July 21, 1998) Supraliminal Method of Education

D'Alitalia, Joseph A., et al. Abstract --- A method of behavior modification involves having a patient view supraliminal video messages superimposed upon an underlying video presentation. The video messages incorporate messages wherein at least some of the messages link a desired modified behavior to positive feelings of the patient. A supraliminal message generator and superimposer iteratively selects individual messages for display from the sequence of messages, decompressing the messages as required, and places the selected messages in a buffer memory of a video generation device. A processor of the supraliminal message generator and superimposer then fades the selected message from an invisible level to a visible level on the video display, and then fades the selected message from the visible level back to the invisible level.

US Patent # 5,649,061 (July 15, 1997) Device and Method for Estimating a Mental Decision

Smyth, Christopher Abstract --- A device and method for estimating a mental decision to select a visual cue from the viewer's eye fixation and corresponding single event evoked cerebral potential. The device comprises an eye tracker, an electronic bio-signal processor and a digital computer. The eye tracker determines the instantaneous viewing direction from oculometric measurements and a head position and orientation sensor. The electronic processor continually estimates the cerebral electroencephalogramic potential from scalp surface measurements following corrections for electrooculogramic, electromyogramic and electro-cardiogramic artifacts. The digital computer analyzes the viewing direction data for a fixation and then extracts the corresponding single event evoked cerebral potential. The fixation properties, such as duration, start and end pupil sizes, end state (saccade or blink) and gaze fixation count, and the parametric representation of the evoked potential are all inputs to an artificial neural network for outputting an estimate of the selection interest in the gaze point of regard. The artificial neural network is trained off-line prior to application to represent the mental decisions of the viewer. The device can be used to control computerized machinery from a video display by ocular gaze point of regard alone, by determining which visual cue the viewer is looking at and then using the estimation of the task-related selection as a selector switch.

US Patent # 5,644,363 (July 1, 1997) Apparatus for Superimposing Visual Subliminal Instructions on a Video Signal

Mead, Talbert Abstract --- A subliminal video instructional device comprises circuitry for receiving an underlying video signal and presenting this signal to horizontal and vertical synchronization detection circuits, circuitry for generating a subliminal video message synchronized to the underlying video signal, and circuitry for adding the subliminal video message to the underlying video signal to create a combination video signal.

US Patent # 5,586,967 (December 24, 1996) Method & Recording for Producing Sounds and Messages to Achieve Alpha & Theta Brainwave States

Davis, Mark E. Abstract --- A method and recording for the use in achieving alpha and theta brainwave states and effecting positive emotional states in humans, is provided which includes a medium having a musical composition thereon with an initial tempo decreasing to a final tempo and verbal phrases recorded in synchrony with the decreasing tempo.

US Patent # 5,562,597 (October 8, 1996) Method & Apparatus for Reducing Physiological Stress

Van Dick, Robert C., Abstract --- Physiological stress in a human subject is treated by generating a weak electromagnetic field about a quartz crystal. The crystal is stimulated by applying electrical pulses of pulse widths between 0.1 and 50 microseconds each at a pulse repetition rate of between 0.5K and 10K pulses per second to a conductor positioned adjacent to the quartz crystal thereby generating a weak electromagnetic field. A subject is positioned within the weak electromagnetic field for a period of time sufficient to reduce stress.

US Patent # 5,551,879 (September 3, 1996) Dream State Teaching Machine

Raynie, Arthur D. Abstract --- A device for enhancing lucidity in the dream state of an individual. The device includes electronic circuitry incorporated into a headband for the user to wear while sleeping. The circuitry includes a detector for fitting adjacent to the eye of the sleeping individual, for detecting Rapid Eye Movement (REM), which occurs during the dream state. The detector emits a signal that is evaluated by additional circuitry to determine whether or not REM sleep is occurring. If REM sleep is occurring, a signal is generated to operate a recorded, which typically plays prerecorded messages through the headphones engaging the ear of the sleeping individual.

US Patent # 5,539,705 (July 23, 1996) Ultrasonic Speech Translator and Communication System

M. A. Akerman, M., et al. Abstract --- A wireless communication system, undetectable by radio-frequency methods, for converting audio signals, including human voice, to electronic signals in the ultrasonic frequency range, transmitting the ultrasonic signal by way of acoustic pressure waves across a carrier medium, including gases, liquids and solids, and reconverting the ultrasonic acoustic pressure waves back to the original audio signal. This invention was made with government support under Contract DE ACO5 840R21400, awarded by the US Department of Energy to Martin Marietta Energy Systems, Inc.

US Patent # 5,507,291 (April 16, 1996) Method & Apparatus for Remotely Determining Information as to Person's Emotional State

Stirbl, et al. Abstract --- In a method for remotely determining information relating to a person's emotional state, a waveform energy having a predetermined frequency and a predetermined intensity is generated and wirelessly transmitted towards a remotely located subject. Waveform energy emitted from the subject is detected and automatically analyzed to derive information relating to the individual's emotional state. Physiological or physical parameters of blood pressure, pulse rate, pupil size, respiration rate and perspiration level are measured and compared with reference values to provide information utilizable in evaluating interviewee's responses or possibly criminal intent in security sensitive areas.

US Patent # 5,522,386 (June 4, 1996) Apparatus for Determination of the Condition of the Vegetative Part of the Nervous System

Lerner, Eduard Abstract --- Apparatus for use in the determination of the condition of the vegetative part of the nervous system and/or of sensory functions of an organism, i.e. a human being or animal. The apparatus comprises devices for generating and supplying to said organism at least one sensory stimulus chosen from a group of sensory stimuli, such as visual, sound, olfactory, gustatory, tactile or pain stimuli, and devices for measuring the skin potential and the evoked response of the organism to a stimulus. The measured data are processed by processing devices for automatically controlling the supply of at least one stimulus for providing a non-rhythmical sequence of stimuli. Preferably, pairs of stimuli are supplied for developing a conditioned reflex.

US Patent # 5,480,374 (January 2, 1996) Method and Apparatus for Reducing Physiological Stress

Van Dick, Robert Abstract --- Physiological stress in a human subject is treated by generating a weak electromagnetic field about a grounded electrode by the application of pulses of between 5 and 50 microseconds each at a pulse rate of between 0.5K and 10K pulses per second to a power electrode, the power electrode and grounded electrode being coupled to high voltage pulse generation means. A subject is positioned within the weak electromagnetic field for a period of time sufficient to cause an increase in his or her alpha or theta brain wave levels.

US Patent # 5,479,941 (January 2, 1996) Device for Inducing Altered States of Consciousness

Harner, Michael Abstract --- A rotating device for producing altered states of consciousness in a subject is provided. The subject's body rotates about a point in the center of the body support means at a speed between about 10 and about 60 revolutions per minute. In a preferred embodiment, the direction of rotation is periodically reversed.

US Patent # 5,392,788 (February 28, 1995) Method and Device for Interpreting Concepts and Conceptual Thought

Hudspeth, William J. Abstract --- A system for acquisition and decoding of EP and SP signals is provided which comprises a transducer for presenting stimuli to a subject, EEG transducers for recording brainwave signals from the subject, a computer for controlling and synchronizing stimuli presented to the subject and for concurrently recording brainwave signals, and either interpreting signals using a model for conceptual perceptional and emotional thought to correspond EEG signals to thought of the subject or comparing signals to normative EEG signals from a normative population to diagnose and locate the origin of brain dysfunctional underlying perception, conception, and emotion.

US Patent # 5,356,368 (October 18, 1994) Method & Apparatus for Inducing Desired States of Consciousness

Monroe, Robert E. Abstract --- Improved methods and apparatus for entraining human brain patterns, employing frequency following response (FFR) techniques, facilitate attainment of desired states of consciousness. In one embodiment, a plurality of electroencephalogram (EEG) waveforms, characteristic of a given state of consciousness, are combined to yield an EEG waveform to which subjects may be susceptible more readily. In another embodiment, sleep patterns are reproduced based on observed brain patterns during portions of a sleep cycle; entrainment principles are applied to induce sleep. In yet another embodiment, entrainment principles are applied in the work environment, to induce and maintain a desired level of consciousness. A portable device also is described

US Patent # 5,352,181 (October 4, 1994) Method & Recording for Producing Sounds and Messages

Davis, Mark E. Abstract --- A method and recording for use in achieving Alpha and Theta brain wave states and effecting positive emotional states in humans to enhance learning and self-improvement, is provided which includes a medium having a musical composition recorded thereon with an initial tempo decreasing to a final tempo and verbal phrases, comprising between approximately 4 and approximately 8 words, recorded in synchrony with the decreasing initial tempo.

US Patent # 5,330,414 (July 19, 1994) Brain Wave Inducing Apparatus

Yasushi, Mitsuo Abstract --- A random signal generator outputs a random noise signal to a band pass filter which selectively passes frequency components in the frequency range of a desired brain wave from a subject. The output of the band pass filter is supplied to an automatic level controller. The automatic level controller sets the output of band pass filter to a predetermined amplitude. Then, the output of the automatic level controller is fed to a stimulating light generator, which converts the output of the automatic level controller into a light signal for stimulating the subject in order to induce the desired brain wave from the subject. The light signal is then emitted into the subject's eyes.

US Patent # 5,289,438 (February 22, 1994) Method & System for Altering Consciousness

Gall, James Abstract --- A system for altering the states of human consciousness involves the simultaneous application of multiple stimuli, preferable sounds, having differing frequencies and wave forms. The relationship between the frequencies of the several stimuli is exhibited by the equation $g = 2.\sup.n/4 \ .multidot.f$ where: f = frequency of one stimulus; g = frequency of the other stimuli or stimulus; and n = a positive or negative integer which is different for each other stimulus.

US Patent # 5,245,666 (September 14, 1993) Personal Subliminal Messaging System

Mikell, Bruce T. Abstract --- A personal subliminal messaging system includes a wide range linear subliminal modulator (43), a digital audio recording or play device (46), a microphone (51) to pick up the sound at the ear, and an earpiece (50) to deliver the subliminal message. The sound level at the user's ear is detected and measured. After rise time and decay conditioning of the varying dc control signal, the wide range linear modulator (43) uses this signal to control the level of the message to the earpiece (50). The user adjusts the system for a liminal of a subliminal level. The psychoacoustic phenomena of Post Masking are used to increase the integrity of the message in subliminal messaging systems.

US Patent # 5,270,800 (December 14, 1993) Subliminal Message Generator

Sweet, Robert L. Abstract --- A combined subliminal and supraliminal message generator for use with a television receiver permits complete control of subliminal messages and their manner of presentation. A video synchronization detector enables a video display generator to generate a video message signal corresponding to a received alphanumeric text message in synchronism with a received television signal. A video mixer selects either the received video signal or the video message signal for output. The messages produced by the video message generator are user selectable via a keyboard input. A message memory stores a plurality of alphanumeric text messages specified by user commands for use as subliminal messages. This message memory preferably includes a read only memory storing predetermined sets of alphanumeric text messages directed to differing topics. The sets of predetermined alphanumeric text messages preferably include several positive affirmations directed to the left brain and an equal number of positive affirmations directed to the right brain that are alternately presented subliminally. The left-brain messages are presented in a linear text mode, while the right brain messages are presented in a three-dimensional perspective mode. The user can control the length and spacing of the subliminal presentations to accommodate differing conscious thresholds. Alternative embodiments include a combined cable television converter and subliminal message generator, a combine television receiver and subliminal message generator and a computer capable of presenting subliminal messages.

US Patent # 5,224,864 (July 6, 1993) Method of Recording and Reproducing Subliminal Signals that are 180 Degrees Out of Phase

Woith, Blake F. Abstract --- A subliminal recording includes both subliminal message and mask signals applied to both tracks of a two-track recording medium. The subliminal message signals are identical in content, and are recorded in an out-of-phase relationship. The mask signals are recorded in phase. The resulting recording may be utilized in the conventional manner for subliminal recordings. By combining the composite signals in an inverted relationship, the mask signals cancel while the subliminal message signals are additive, thus allowing the presence of the subliminal message signal to be confirmed on the recording.

US Patent # 5,221,962 (June 22, 1993) Subliminal Device having Manual Adjustment of Perception Level of Subliminal Messages

Backus, Alan L., et al. Abstract --- A method and apparatus for presenting subliminal visual and/or audio messages which allows user verification of message content and presence, as well as proper adjustment of message obviousness while accounting for ambient conditions and user sensitivities is disclosed. This method and apparatus also presents synchronized reinforced sensory input of subliminal messages. This is performed by simultaneously overlaying images received from a VCR over a plurality of television signals. This apparatus directs overlay images over RF television signals having both audio and video components

US Patent # 5,215,468 (June 1, 1993) Method and Apparatus for Introducing Subliminal Changes to Audio Stimuli

Lauffer, Martha A., et al. Abstract --- A method and apparatus for introducing gradual changes to an audio signal so that the changes are subliminal. The changes can involve tempo and volume, for example, and can take the form of a gentle gradient having ever increasing/decreasing ramp-like changes over a sufficient duration, or a more complex program involving several gentle gradients. In the preferred embodiment, an enhanced audio play-back device such as a portable audio cassette recorder can be programmed to subliminally alter the characteristics of a standard pre-recorded tape containing music, for example. As a motivational tool during walking, jogging or other repetitive exercise, the tempo is gradually increased over a period of time to encourage a corresponding gradual (and subliminal) increase in physical exertion by a user whose rate of movement is proportional to the tempo of the music. The tempo can be either manually changed in conjunction with a subliminal program, or by itself in an override mode, or by itself in a version of the present-inventive audio play-back device which allows only manual tempo alternation. In an alternate embodiment, a special pre-recorded tape contains subliminal changes in tempo, for example, for play-back on a standard audio cassette recorder (which operates at one speed, only) to cause the same effect as the preferred embodiment.

US Patent # 5,213,562 (May 25, 1993) Method of Inducing Mental, Emotional and Physical States of Consciousness

Monroe, Robert A. Abstract --- A method having applicability in replication of desired consciousness states; in the training of an individual to replicate such a state of consciousness without further audio stimulation; and in the transferring of such states from one human being to another through the imposition of one individual's EEG, superimposed on desired stereo signals, on another individual, by inducement of a binaural beat phenomenon.

US Patent # 5,194,008 (March 16, 1993) Subliminal Image Modulation Projection and Detection System and Method

Mohan, William L., et al. Abstract --- Weapon training simulation system including a computer operated video display scene whereon is projected a plurality of visual targets. The computer controls the display scene and the targets, whether stationary or moving, and processes data of a point of aim sensor apparatus associated with a weapon operated by a trainee. The sensor apparatus is sensitive to non-visible or subliminal modulated areas having a controlled contrast of brightness between the target scene and the targets. The sensor apparatus locates a specific subliminal modulated area and the computer determines the location of a target image on the display scene with respect to the sensor apparatus

US Patent # 5,175,571 (December 29, 1992) Glasses with Subliminal Message

Tanefsky, Faye, et al. Abstract --- A pair of subliminal imaging spectacles is provided with a matched pair of visual subliminal images designed and placed so as to merge into one image due to the stereoscopic effect of human vision and thus to impart a subliminal message to the wearer.

US Patent # 5,170,381 (December 8, 1992) Method for Mixing Audio Subliminal Recordings

Taylor, Eldon, et al. Abstract --- Audio subliminal recordings are made in which in addition to using a primary carrier, such as music, two audio channels are used to deliver subliminal messages to the brain. On one channel, accessing the left brain hemisphere, the message delivered is meaningfully spoken, forward-masked, permissive affirmations delivered in a round-robin manner by a male voice, a female voice and a child's voice. On the other channel, accessing the right brain, directive messages, in the same voices, are recorded in backward-masked (or meta-contrast). The three voices are recording in round-robin fashion with full echo reverberation. The audio tracks are mixed using a special processor which converts sound frequencies to electrical impulses and tracks the subliminal message to synchronize the subliminal message in stereo with the primary carrier. The processor maintains constant gain differential between the primary carrier and the subliminal verbiage and, with the subliminal verbiage being recorded with round-robin, full echo reverberation, ensures that none of a message is lost. The primary carrier should be continuous music without breaks or great differences in movements.

US Patent # 5,159,703 (October 27, 1992) Silent Subliminal Presentation System

Lowery, Oliver Abstract --- A silent communications system in which non-aural carriers, in the very low or very high audio frequency range or in the adjacent ultrasonic frequency spectrum, are amplitude or frequency modulated with the desired intelligence and propagated acoustically or vibrationally, for inducement into the brain, typically through the use of loudspeakers, earphones or piezoelectric transducers.

US Patent # 5,151,080 (September 29, 1992) Method & Apparatus for Inducing & Establishing a Changed State of Consciousness

Bick, Claus Abstract --- An electroacoustic device includes a sound generator as well as a system for producing synthetic human speech, connected to a modulation stage for superimposing the output signals thereof. The superimposed output signals are applied via an amplifier stage to one of a headphone system or loudspeaker system.

US Patent # 5,135,468 (August 4, 1992) Method & Apparatus of Varying the Brain State of a Person by Means of an Audio Signal

Meissner, Juergen P. Abstract --- A method of varying the brain state of a person includes the steps of supplying the first audio signal to one ear of the person, supplying a second audio signal to the other ear of the person, and substantially continuously varying the frequency of at least one of the first and second audio signals to vary the brain state of the person

US Patent # 5,134,484 (July 28, 1992) Superimposing Method & Apparatus Useful for Subliminal Messages

Wilson, Joseph Abstract --- Data to be displayed is combined with a composite video signal. The data is stored in a memory in digital form. Each byte of the data is read out in sequential fashion to determine: the recurrence display rate of the data according to the frame sync pulses of the video signal; the location of the data within the video image according to the line sync pulses of the video signal; and the location of the data display within the video image according to the position information. Synchronization of the data with the video image is derived from the sync pulses of the composite video signal. A similar technique is employed to combine sound data with an audio signal. Data to be displayed may be presented as a subliminal message or may persist for a given time interval. The data may be derived from a variety of sources including a prerecorded or live video signal. The message may be a reminder message displayed upon a television screen to remind the viewer of an appointment. The data may be stored in a variety of different memory devices capable of high speed data retrieval. The data may be generated locally on-line or off-line and transferred to memory which stores the data necessary to create the message.

US Patent # 5,128,765 (July 7, 1992) System for Implementing the Synchronized Superimposition of Subliminal Signals

Dingwall, Robert Abstract --- An apparatus and system for the controlled delivery of a subliminal video and/or audio message on to a source signal from a video tape player or similar. The source signal is divided into audio and video portions. A video processor reads synchronization information from the source signal. A controller transmits a stored subliminal image at designated times to a mixer amplifier fully synchronized with the source signal. Concurrently, an audio subliminal message is applied to the source audio at a volume level regulated at some fraction to the source audio. The combined signals are transmitted to a monitor for undistracted viewing.

US Patent # 5,123,899 (June 23, 1992) Method & System for Altering Consciousness

Gall, James Abstract --- A system for altering the states of human consciousness involves the simultaneous application of multiple stimuli, preferable sounds, having differing frequencies and wave forms. The relationship between the frequencies of the several stimuli is exhibited by the equation $g = s.sup.n/4 \;.multidot.f$ where: f = frequency of one stimulus; g = frequency of the other stimuli of stimulus; and $n=a$ positive or negative integer which is different for each other stimulus.

US Patent # 5,052,401 (October 1, 1991) Sherwin, Gary Product Detector for a Steady Visual Evoked Potential Stimulator and Product Detector

Abstract --- An automated visual testing system is disclosed which presents an alternating steady state visual stimulus to a patient through an optical system that modifies the stimulus image. As the image changes, the patient produces evoked potentials that change. The evoked potentials are detected by a product detector which produces the amplitude of the evoked potentials. The product detector includes filters which isolate the patient's evoked potentials, a modulator which detects the response using the stimulus source frequency and a demodulator that determines the amplitude of the response. The product detector detects the level of the steady state evoked potential signals even in the presence of substantial background noise and extraneous electroencephalographic signals. These detectors can be used to monitor the evoked potential produced by visual, aural or somatic steady state stimuli. The components described above can be used to produce a system that can determine to which of several different displays an observer is paying attention by providing images that blink at different frequencies and product detectors for each of the stimulus frequencies. The product detector producing the highest output indicates the display upon which the observer is focused.

US Patent # 5,047,994 (September 10, 1991) Supersonic Bone Conduction Hearing Aid and Method

Lenhardt, Martin., et al. Abstract --- A supersonic bone conduction hearing aid that receives conventional audiometric frequencies and converts them to supersonic frequencies for connection to the human sensory system by vibration bone conduction. The hearing is believed to use channels of communications to the brain that are not normally used for hearing. These alternative channels do not deteriorate significantly with age as does the normal hearing channels. The supersonic bone conduction frequencies are discerned as frequencies in the audiometric range of frequencies.

US Patent # 5,036,858 (August 6, 1991) Method & Apparatus for Changing Brain Wave Frequency

Carter, John L., et al. Abstract --- A method for changing brain wave frequency to a desired frequency determines a current brain wave frequency of a user, generates two frequencies with a frequency difference of a magnitude between that of the current actual brain wave frequency and the desired frequency but always within a predetermined range of the current actual brain wave frequency, and produces an output to the user corresponding to the two frequencies. One apparatus to accomplish the method has a computer processor, a computer memory, EEG electrodes along with an amplifier, a programmable timing generator responsive to the computer processor for generating the two frequencies, audio amplifiers and a beat frequency generator driving a visual frequency amplifier.

US Patent # 5,027,208 (June 25,1991) Therapeutic Subliminal Imaging System

Dwyer, Jr., Joseph, et al. Abstract --- A therapeutic subliminal imaging system wherein a selected subliminal message is synchronized with and added to an existing video signal containing a supraliminal message. A television receiver or video recorder can be used to provide the supraliminal message and a video processing circuit varies the intensity of that perceptible message to incorporate one or more subliminal images.

US Patent # 5,017,143 (May 21, 1991) Method and Apparatus for Producing Subliminal Images

Backus, Alan, et al. Abstract --- A method and apparatus to produce more effective visual subliminal communications. Graphic and/or text images, presented for durations of less than a video frame, at organized rhythmic intervals, the rhythmic intervals intended to affect user receptivity, moods or behavior. Subliminal graphic images having translucent visual values locally dependent on background values in order to maintain desired levels of visual contrast.

US Patent # 4,958,638 (September 25, 1990) Non-Contact Vital Signs Monitor

Sharpe, Steven, et al. Abstract --- An apparatus for measuring simultaneous physiological parameters such as heart rate and respiration without physically connecting electrodes or other sensors to the body. A beam of frequency modulated continuous wave radio frequency energy is directed towards the body of a subject. The reflected signal contains phase information representing the movement of the surface of the body, from which respiration and heartbeat information can be obtained. The reflected phase modulated energy is received and demodulated by the apparatus using synchronous quadrature detection. The quadrature signals so obtained are then signal processed to obtain the heartbeat and respiratory information of interest.

US Patent # 4,924,744 (May 15, 1990) Apparatus for Generating Sound through Low Frequency and Noise Modulation

Lenzen, Reiner Abstract --- In an apparatus for generating sound, there are provided a plurality of channels for generating sounds. Each of the channels includes a memory for storing waveform data, and at least one of the channels includes a noise generator so that various kinds of sounds including rhythm sound-effects sound, effects sound-vibrato etc. are generated. There is further provided a controller by which voice sound signal is passed through the channels so that artificial sound, voice sound etc. are generated. There is still further provided a circuit for adjusting an amplitude level of a whole sound which is obtained by mixing output sounds of the channels so that far and near sound is produced. Further, each of the channels includes left and right attenuators which divide a channel sound into left and right channel sounds. Still further, the apparatus comprises a low frequency oscillator for controlling a depth of frequency modulation, and a controller for writing sampling data of a predetermined waveform into serial addresses of a memory.

US Patent # 4,889,526 (December 26, 1989) Non-Invasive Method & Apparatus for Modulating Brain Signals

Rauscher, Elizabeth A. Abstract --- This invention incorporates the discovery of new principles which utilize magnetic and electric fields generated by time varying square wave currents of precise repetition, width, shape and magnitude to move through coils and cutaneously applied conductive electrodes in order to stimulate the nervous system and reduce pain in humans. Timer means, adjustment means, and means to deliver current to the coils and conductive electrodes are described, as well as a theoretical model of the process. The invention incorporates the concept of two cyclic expanding and collapsing magnetic fields which generate precise wave forms in conjunction with each other to create a beat frequency which in turn causes the ion flow in the nervous system of the human body to be efficiently moved along the nerve path where the locus of the pain exists to thereby reduce the pain. The wave forms are created either in one or more coils, one or more pairs of electrodes, or a combination of the two.

US Patent # 4,883,067 (November 28, 1989) Method & Apparatus for Translating the EEG into Music

Knispel, Joel, et al. Abstract --- A method and apparatus for applying a musical feedback signal to the human brain, or any other brain, to induce controllable psychological and physiological responses. A signal representing the ongoing electroencephalographic (EEG) signal of a brain preferably is obtained from the electrode location on the scalp known as CZ or P3 in clinical notation. A signal processor converts the ongoing EEG into electrical signals which are converted into music by synthesizers. The music is acoustically fed back to the brain after a time delay calculated to shift the phase of the feedback in order to reinforce specific or desired ongoing EEG activity from the scalp position of interest. The music is comprised of at least one voice that follows the moment-by-moment contour of the EEG in real time to reinforce the desired EEG activity. The music drives the brain into resonance with the music to provide a closed loop or physiological feedback effect. Preferably, the musical feedback comprises additional voices that embody psychoacoustic principles as well as provide the content and direction normally supplied by the therapist in conventional biofeedback. The invention contemplates numerous applications for the results obtained.

US Patent # 4,877,027 (October 31, 1989) Hearing System

Brunkan, Wayne B. Abstract --- Sound is induced in the head of a person by radiating the head with microwaves in the range of 100 megahertz to 10,000 megahertz that are modulated with a particular waveform. The waveform consists of frequency modulated bursts each burst is made up of 10 to 20 uniformly spaced pulses grouped tightly together the burst width is between 500 nanoseconds and 100 microseconds. The pulse width is in the range of 10 nanoseconds to 1 microsecond. The bursts are frequency modulated by the audio input to create the sensation of hearing in the person whose head is irradiated.

US Patent # 4,858,612 (August 22, 1989) Hearing Device

Stocklin, Philip L. Abstract --- A method and apparatus for stimulation of hearing in mammals by introduction of a plurality of microwaves into the region of the auditory cortex is shown and described. A microphone is used to transform sound signals into electrical signals which are in turn analyzed and processed to provide controls for generating a plurality of microwave signals at different frequencies. the multi-frequency microwaves are then applied to the brain in the region of the auditory cortex. By this method sounds are perceived by the manual which are representative of the original sound received by the microphone.

US Patent # 4,834,701 (May 30, 1989) Apparatus for Inducing Frequency Reduction in Brain Wave

Masaki, Kazumi Abstract --- Frequency reduction in human brain wave is inducible by allowing human brain to perceive 4-16 hertz beat sound. Such beat sound can be easily produced with an apparatus, comprising at least one sound source generating a set of low-frequency signals different each other in frequency by 4-16 hertz.

Electroencephalographic study revealed that the beat sound is effective to reduce beta-rhythm into alpha-rhythm, as well as to retain alpha-rhythm.

US Patent # 4,821,326 (April 11, 1989) Non-Audible Speech Generation Method & Apparatus

MacLeod, Norman Abstract --- A non-audible speech generation apparatus and method for producing non-audible speech signals which includes an ultrasonic transducer or vibrator for projecting a series of glottal shaped ultrasonic pulses to the vocal track of a speaker. The glottal pulses, in the approximate frequency spectrum extending from 15 kilohertz to 105 kilohertz, contains harmonics of approximately 30 times the frequency of the acoustical harmonics generated by the vocal cords, but which may nevertheless be amplitude modulated to produce non-audible speech by the speaker's silently mouthing of words. The ultrasonic speech is then received by an ultrasonic transducer disposed outside of the speaker's mouth and electronically communicated to a translation device which down converts the ultrasonic signals to corresponding signals in the audible frequency range and synthesizes the signals into artificial speech.

US Patent # 4,777,529 (October 11, 1988) Auditory Subliminal Programming System

Schultz, Richard M., et al. Abstract --- An auditory subliminal programming system includes a subliminal message encoder that generates fixed frequency security tones and combines them with a subliminal message signal to produce an encoded subliminal message signal which is recorded on audio tape or the like. A corresponding subliminal decoder/mixer is connected as part of a user's conventional stereo system and receives as inputs an audio program selected by the user and the encoded subliminal message. The decoder/mixer filters the security tones, if present, from the subliminal message and combines the message signals with selected low frequency signals associated with enhanced relaxation and concentration to produce a composite auditory subliminal signal. The decoder/mixer combines the composite subliminal signal with the selected audio program signals to form composite signals only if it detects the presence of the security tones in the subliminal message signal. The decoder/mixer outputs the composite signal to the audio inputs of a conventional audio amplifier where it is amplified and broadcast by conventional audio speakers.

US Patent # 4,734,037 (March 29, 1988) Message Screen

McClure, J. Patrick Abstract --- A transparent sheet is disclosed having a message thereon. The sheet has a first side adapted to be attached facing a plate which is normally viewed by a viewer and a second side facing the viewer. The message is arranged to be readably intelligible from the second side but is not visible to the viewer when viewed from a normal viewing distance from the second side under normal viewing conditions. The message has a subliminal effect upon the viewer when viewed from the normal viewing distance from the second side under normal viewing conditions. A viewer can electively subject him or herself to subliminal messages while viewing television at leisure.

US Patent # 4,717,343 (January 5, 1988) Method of Changing a Person's Behavior

Densky, Alan B. Abstract --- A method of conditioning a person's unconscious mind in order to effect a desired change in the person's behavior which does not require the services of a trained therapist. Instead the person to be treated views a program of video pictures appearing on a screen. The program as viewed by the person's unconscious mind acts to condition the person's thought patterns in a manner which alters that person's behavior in a positive way.

US Patent # 4,699,153 (October 13, 1987) System for Assessing Verbal Psychobiological

Correlates Shevrin, Howard, et al. Abstract --- A system for assessing psychobiological conditions of a subject utilizes a plurality of words which are selected to be in four categories as critical stimuli. The words are presented by a tachistoscope to the subject in subliminal and supraliminal modes of operation. Subliminal stimulation of the subject is achieved by presenting the selected words for an exposure period of approximately one millisecond. The supraliminal exposure time is approximately thirty milliseconds. Prior to stimulation, the subject is diagnosed in accordance with conventional psychoanalytical techniques to establish the presence and nature of a pathological condition. The words are selected and categorized in four groups: pleasant words, unpleasant words, words related to a diagnosed conscious pathological condition, and words related to a diagnosed unconscious pathological condition. The brain wave responses which are evoked by the stimulation are collected via electrodes and analyzed in accordance with a trans-information technique which is based on information signal theory for establishing a probabilistic value which corresponds to the information content of the evoked responses.

US Patent # 4,692,118 (September 8, 1987) Video Subconscious Display Attachment

Mould, Richard E. Abstract --- An apparatus and method for introducing messages to the subconscious mind is disclosed, which includes a panel positioned adjacent a television screen, with the panel having non-distractive messages imprinted thereon, such that as the subject consciously focuses his attention on the video screen, his subconscious mind records the message from the panel that is within his peripheral vision.

US Patent # 4,616,261 (October 7, 1986) Method & Apparatus for Generating Subliminal Visual Messages

Crawford, James R., et al. Abstract --- A system for generating a subliminal message during the display of a normal television program on a television receiver utilizes a personal computer to generate an RF carrier modulated with video signals encoding the subliminal message. The computer runs under the control of an application program which stores the subliminal message and also controls the computer to cause it to generate timing signals that are provided to a single pole double-throw switch. The source of the normal television program and the video output of the computer are connected to the two switch inputs and the switch output is connected to the television receiver antenna system. The timing signals cause the switch to normally display the conventional television program and to periodically switch to the computer output to generate the subliminal message. The video output of the computer includes horizontal and vertical synchronizing signals which are of substantially the same frequency as the synchronizing signals incorporated within the normal program source but of an arbitrary phase.

US Patent # 4,573,449 (March 4, 1986) Method for Stimulating the Falling Asleep and/or Relaxing Behavior of a Person

Warnke, Egon F., Abstract --- A method and apparatus is provided with which a person suffering from sleeplessness can be more easily relaxed and may more rapidly fall asleep. In particular, sound pulses are emitted by an electro-acoustic transducer, according to the cadence of which, the person seeking to fall asleep is induced to breathe in and out over a predetermined period of time. By suitably selecting the pulse sequence frequency, the pitch and the amplitude of the sound pulses may be adjusted thereby enhancing the process of falling asleep.

US Patent # 4,508,105 (April 2, 1985) Shadow Generating Apparatus

Whitten, Glen, et al. Abstract --- Disclosed is an apparatus for inducing various brain wave patterns through visual stimulation. The apparatus comprises a pair of spectacles or other viewing apparatus having a liquid crystal display embedded in each lens. By repetitively activating and deactivating the liquid crystals, shadows are generated which are perceived by the subject individual wearing the viewing apparatus. Responding to the frequency of shadow generation, the subject's brain is thereby induced to generate sympathetic brain wave frequencies. The apparatus finds particular utility in the generation of alpha waves. Because learning is enhanced when the brain is in the alpha state, activities such as listening to tapes or lectures and the like can be carried out with greater facility. Shadow generation is accomplished through the use of a timing mechanism for each liquid crystal display and the frequency for each is adjustable over a wide range, permitting synchronous or asynchronous timing.

US Patent # 4,395,600 (July 26, 1983) Auditory Subliminal Message System & Method

Lundy, Rene R., et al. Abstract --- Ambient audio signals from the customer shopping area within a store are sensed and fed to a signal processing circuit that produces a control signal which varies with variations in the amplitude of the sensed audio signals. A control circuit adjusts the amplitude of an auditory subliminal anti-shoplifting message to increase with increasing amplitudes of sensed audio signals and decrease with decreasing amplitudes of sensed audio signals. This amplitude controlled subliminal message may be mixed with background music and transmitted to the shopping area. To reduce distortion of the subliminal message, its amplitude is controlled to increase at a first rate slower than the rate of increase of the amplitude of ambient audio signals from the area. Also, the amplitude of the subliminal message is controlled to decrease at a second rate faster than the first rate with decreasing ambient audio signal amplitudes to minimize the possibility of the subliminal message becoming supraliminal upon rapid declines in ambient audio signal amplitudes in the area. A masking signal is provided with an amplitude which is also controlled in response to the amplitude of sensed ambient audio signals. This masking signal may be combined with the auditory subliminal message to provide a composite signal fed to, and controlled by, the control circuit.

US Patent # 4,388,918 (June 21, 1983) Mental Harmonization Process

Filley, Charles C. Abstract --- A state of relaxation or mental harmonization in a subject is created by exposing a color solely to one field of vision of a subject and the complement of that color solely to the other field of vision of the subject while simultaneously exposing an audible tone solely to one ear of the subject and a harmonious tone solely to the other ear of the subject. The color and tones employed are subjectively comfortable and compatible. Preferably, the frequency difference between the two audible tones is one-half the frequency of the audible tone having the lowest frequency.

US Patent # 4,354,505 (October 19, 1982) Method of and Apparatus for Testing and Indicating Relaxation State of a Human Subject

Shiga, Kazumasa, Abstract --- In a self-training biofeedback system, a physiological signal representing the state of relaxation of a person using the system is applied to a time counter to generate a binary count output representing the relaxation period. A visual indicator connected to the time counter provides the self-trained person with a quick display of the measured time period so he can gauge the depth of his relaxation.

US Patent # 4,335,710 (June 22, 1982) Device for the Induction of Specific Brain Wave Patterns

Williamson, John Abstract --- Brain wave patterns associated with relaxed and meditative states in a subject are gradually induced without deleterious chemical or neurological side effects. A white noise generator (11) has the spectral noise density of its output signal modulated in a manner similar to the brain wave patterns by a switching transistor within a spectrum modulator and converted to an audio signal by acoustic transducer. Ramp generator gradually increases the voltage received by and resultant output frequency of voltage controlled oscillator whereby switching transistor periodically shunts the high frequency components of the white noise signal to ground.

US Patent # 4,315,501 (February 16, 1982) Learning-Relaxation Device

Gorges, Denis E., Abstract --- Disclosed is a device for relaxing, stimulating and/or driving brain wave form function in a human subject. The device comprises, in combination, an eye mask having independently controlled left and right eyepieces and a peripheral light array in each eyepiece, an audio headset having independently controlled left and right earpieces and a control panel which controls light and sound signals to the light arrays and earpieces, respectively. Various control functions allow simultaneous or alternating light and sound pulsations in the left and right light arrays and earpieces, as well as selective phasing between light and sound pulsations.

US Patent # 4,227,516 (October 14, 1980) Apparatus for Electrophysiological Stimulation

Meland, Bruce C., et al. Abstract --- Apparatus for the electrophysiological stimulation of a patient is provided for creating an analgesic condition in the patient to induce sleep, treat psychosomatic disorders, and to aid in the induction of electro hypnosis and altered states of consciousness. The foregoing is achieved by repetitive stimuli in the patient for whom external influences, namely those of sight and sound, are intentionally excluded. The apparatus produces electrical stimulation of the patient in the form of a modulated wave which produces impulses in the delta, theta, alpha and beta regions of the brain's electrical activity, the electrical stimulation being accompanied by two sources of audio stimulation, one of which is a sinusoidal tone modulated by and synchronized with the electrical stimulation, and the other is derived from sound recordings.

US Patent # 4,191,175 (March 4, 1980) Method & Apparatus for Repetitively Producing a Noise-like Audible Signal

Nagle, William L. Abstract --- A digital pulse generator and shift register repetitively produce bursts of digital pulses at a first adjustable repetition frequency. The repetition frequency of the pulses in each burst is also adjustable. A pink noise filter accentuates the lower burst frequency components near 7 hz and substantially attenuates all frequency components of the bursts above a first cut-off point near 10 Khz. A tunable band pass amplifier having a center frequency adjustable over a preselected range of frequencies optimally detectable by the average human ear accentuates the pink noise filter output near 2.6 Khz. The tunable amplifier drives an audible signal source with noise-like pulses of varying amplitudes and frequency components. A low pass amplifier may be connected to the pink noise filter to generate a train of pulses having a repetition frequency near 7 hz which pulses a light source in synchronism with the audible noise-like signal.

US Patent # 4,141,344 (February 27, 1979) Sound Recording System

Barbara, Louis J., Abstract --- In recording an audio program, such as music or voice, on a magnetic tape recorder an A.C. signal generator operating at a frequency below about 14 Hz provides an AC baseline for the audio program signal. This 14 Hz or lower AC signal is sensed by the listener's ear to create an Alpha or Theta state in his brain when the tape is played back.

US Patent # 4,082,918 (April 4, 1978) Audio Analgesic Unit

Chang, Roland W., et al. Abstract --- An audio analgesic unit for use in masking sounds and substituting another sound which includes earmuffs to be used by a dental patient in which speakers are arranged and connected to a patient operated remote control unit to control the sound levels and a master control unit to override the patient remote control unit and operated by an operator, such as a dentist. A beeper indicates operation mode change.

US Patent # 4,034,741 (July 12, 1977) Noise Generator & Transmitter

Adams, Guy E., et al. Abstract --- An analgesic noise generator employs a circuit that can be switched to provide a variable waveform from an active noise source out of an integrated circuit amplifier.

US Patent # 3,967,616 (July 6, 1976) Multichannel System for & Multifactorial Method of Controlling the Nervous System of a Living Organism

Ross, Sidney A., Abstract - A novel method for controlling the nervous system of a living organism for therapeutic and research purposes, among other applications, and an electronic system utilized in, and enabling the practice of, the invented method. Bioelectrical signals generated in specific topological areas of the organism's nervous system, typically areas of the brain, are processed by the invented system so as to produce a sensory stimulus if the system detects the presence or absence, as the case may be, of certain characteristics in the waveform patterns of the bioelectrical signals being monitored. The coincidence of the same or different characteristics in two or more waveform patterns, or the non-coincidence thereof, may be correlated with a certain desired condition of the organism's nervous system; likewise, with respect to the coincidence or non-coincidence of different characteristics of a single waveform pattern. In any event, the sensory stimulus provided by the invented system, typically an audio or visual stimulus, or combination thereof, is fed back to the organism which associates its presence with the goal of achieving the desired condition of its nervous system. Responding to the stimulus, the organism can be trained to control the waveform patterns of the monitored bioelectrical signals and thereby, control its own nervous system. The results of the coincidence function permit results heretofore unobtainable.

US Patent # 3,884,218 (May 20, 1975) Method of Inducing & Maintaining Various Stages of Sleep in the Human Being

Monroe, Robert A., Abstract --- A method of inducing sleep in a human being wherein an audio signal is generated comprising a familiar pleasing repetitive sound modulated by an EEG sleep pattern. The volume of the audio signal is adjusted to overcome the ambient noise and a subject can select a familiar repetitive sound most pleasing to himself.

US Patent # 3,837,331 (September 24, 1974) System & Method for Controlling the Nervous System of a Living Organism

Ross, S., Abstract --- A novel method for controlling the nervous system of a living organism for therapeutic and research purposes, among other applications, and an electronic system utilized in, and enabling the practice of the invented method. Bioelectrical signals generated in specific topological areas of the organism's nervous system, typically areas of the brain, are processed by the invented system so as to produce an output signal which is in some way an analog of selected characteristics detected in the bioelectrical signal. The output of the system, typically an audio or visual signal, is fed back to the organism as a stimulus. Responding to the stimulus, the organism can be trained to control the waveform pattern of the bioelectrical signal generated in its own nervous system.

US Patent # 3,835,833 (September 17, 1974) Method for Obtaining Neurophysiological Effects

Limoge, A., Abstract --- A method and apparatus for obtaining neurophysiological effects on the central and/or peripheral systems of a patient. Electrodes are suitably positioned on the body of the patient and a composite electric signal is applied at the electrodes. The composite signal is formed by the superpositioning of two signals: a first signal which is a rectified high-frequency carrier modulated in amplitude to about 100 percent by substantially square-shaped pulses whose duration, amplitude and frequency are chosen according to the neurophysiological effects desired, and a second signal which has a relatively white noise spectrum. The mean value of the first electric signal has a predetermined sign which is opposite the sign of the mean value of the second electric signal.

US Patent # 3,773,049 (November 20, 1973) Apparatus for Treatment of Neuropsychic & Somatic Diseases with Heat, Light, Sound & VHF Electromagnetic Radiation

L. Y. Rabichev, et al. Abstract --- N/A

US Patent # 3,766,331 (October 16, 1973) Hearing Aid for Producing Sensations in the Brain

Zink, Henry R., Abstract --- A pulsed oscillator or transmitter supplies energy to a pair of insulated electrodes mounted on a person's neck. The transmitter produces pulses of intensity greater than a predetermined threshold value and of a width and rate so as to produce the sensation of hearing without use of the auditory canal, thereby producing a hearing system enabling otherwise deaf people to hear.

US Patent # 3,727,616 (March 17, 1973) Electronic System for Stimulation of Biological Systems

Lenskes, H., Abstract --- A receiver totally implanted within a living body is inductively coupled by two associated receiving coils to a physically unattached external transmitter which transmits two signals of different frequencies to the receiver via two associated transmitting coils. One of the signals from the transmitter provides the implanted receiver with precise control or stimulating signals which are demodulated and processed in a signal processor network in the receiver and then used by the body for stimulation of a nerve, for example, while the other signal provides the receiver with a continuous wave power signal which is rectified in the receiver to provide a source of electrical operating power for the receiver circuitry without need for an implanted battery.

US Patent # 3,712,292 (January 23, 1973) Method & Apparatus for Producing Swept FM Audio Signal Patterns for Inducing Sleep

Zentmeyer, J. Abstract --- A method of producing sound signals for inducing sleep in a human being, and apparatus therefor together with REPRESENTATIONS thereof in recorded form, wherein an audio signal is generated representing a familiar, pleasing, repetitive sound, modulated by continuously sweeping frequencies in two selected frequency ranges having the dominant frequencies which occur in electrical wave patterns of the human brain during certain states of sleep. The volume of the audio signal is adjusted to mask the ambient noise and the subject can select any of several familiar, repetitive sounds most pleasing to him.

US Patent # 3,647,970 (March 7, 1972) Method and System for Simplifying Speech Waveforms

Flanagan, G. Patrick, Abstract --- A complex speech waveform is simplified so that it can be transmitted directly through earth or water as a waveform and understood directly or after amplification.

US Patent # 3,629,521 (January 8, 1970) Hearing Systems

Puharich, Henry K., Abstract --- The present invention relates to the stimulation of the sensation of hearing in persons of impaired hearing abilities or in certain cases persons totally deaf utilizing RF energy. More particularly, the present invention relates to a method and apparatus for imparting synchronous AF or ""acoustic" signals and so-called "transdermal" or RF signals. Hearing and improved speech discrimination, in accordance with one aspect of the present invention, is stimulated by the application of an AF acoustical signal to the "ear system" conventional bio mechanism of hearing, which is delivered to the brain through the "normal" channels of hearing and a separate transdermal RF electrical signal which is applied to the "facial nerve system" and is detectable as a sensation of hearing. Vastly improved and enhanced hearing may be achieved...

US Patent # 3,576,185 (April 27, 1971) Sleep-Inducing Method & Arrangement using Modulated Sound & Light

Meseck, Oscar & Schulz, Hans R. Abstract --- N/A

US Patent # 3,568,347 (February 23, 1971) Psycho-Acoustic Projector

Flanders, Andrew, Abstract --- A system for producing aural psychological disturbances and partial deafness in the enemy during combat situations.

US Patent # 3,393,279 (July 16, 1968) Nervous System Excitation Device

Flanagan, Giles P., Abstract --- A method of transmitting audio information via a radio frequency signal modulated with the audio info through electrodes placed on the subject's skin, causing the sensation of hearing the audio information in the brain.

US Patent # 3,170,993 (February 23, 1965) Means for Aiding Hearing by Electrical Stimulation of the Facial Nerve System

Puharich, Henry & Lawrence, Joseph, Abstract --- N/A

US Patent # 3,156,787 (November 10, 1964) Solid State Hearing System

Lawrence, Joseph & Puharich, Henry Abstract --- N/A

US Patent # 2,995,633 (August 8, 1961) Means for Aiding Hearing

Puharich, Henry & Lawrence, J.

US Patent - 5,629,678 - Implantable Transceiver

US Patent - 5,878,155 – Barcode Tattoo

US Patent 5,539,705 Ultrasonic Speed Translator and Communications System

US Patent 5,629,678 Personal Tracking and Recovery System

US Patent 5,760,692 Intro - Oral Tracking Device

US Patent 5,868,100 Fenceless Animal Control System Using GPS Location Information

US Patent 4,717,343 Method for Changing a Person's Behavior

US Patent 5,270,800 Subliminal Message Generator

US Patent 4,877,027 Hearing System (Microwave Voice to Skull)

US Patent 6,011,991 Communication System and Method Including Brain Wave Analysis and/or use of Brain Activity

US Patent 4,858,612 Hearing Device

U.S. Patent 3,951,134 Apparatus and Method for Remotely Monitoring and Altering Brain Waves

US Patent - 5,507,291 Method and an Associated Apparatus for Remotely Determining Information as to a Person's Emotional State

U.S. Patent 5,905,461 Global Positioning Satellite Tracking Device

U.S. Patent 5,935,034 Magnetic Excitation and Sensory Resonances

U.S. Patent 5,952,600 Engine Disabling Weapon

U.S. Patent 6,006,188 Speech Signal Processing for Determining Psychological or Physiological Characteristics Using a Knowledge Base

U.S. Patent 6,014,080 Body Worn Active and Passive Tracking Device

U.S. Patent 6,017,302 Subliminal Acoustic Manipulation of the Nervous System

US Patent 6,052,336 Apparatus and Method of Broadcasting Audible Sound Using Ultrasound as a Carrier

Keep in mind that the bulk of these technologies are now highly perfected through years and years of perfection. They are sophisticated and are available in various forms, to include delivery via satellite. Some are said to be capable of delivery through H.A.A.R.P from the remote location in Alaska and delivery systems by communication towers are strategically placed around the world through GWEN communication towers. Today, there are patents capable of doing anything the mind of man can imagine and there are numerous patents for subliminal message caring technology.

Science has decoded the brain signals so that direct communication with the brain is possible. The brain's electromagnetic signals can be remotely detected by SQUID. SQUIDs (superconducting quantum interference devices) have military applications in magnetometers used to detect submarines and mines. SQUIDs can detect the most tenuous magnetic fields, even those generated by brain cells. They exploit the properties of a Josephson junction, which is constructed from two superconductors separated by a non-superconducting layer. Lockheed has built an HTSC SQUID. The extreme sensitivity of SQUIDs makes them ideal for studies in biology.

Magneto encephalography (MEG) for example uses measurements from an array of SQUIDs to make inferences about neural activity inside brains. SQUID has the capability of gaging the target's emotional state using neuro-feedback from the central nervous system via satellite radar. If a target's emotional state is accurately assessed, then those tracking the individual can use various tactics in a subliminal manipulative effort or other forms of mind control.

DARPA set up the Total Information Awareness program in 2002, which promised to provide "total information awareness" through "large, distributed repositories" including "biometric signatures of humans" and "human network analysis and behavior modeling. "Biometric satellite surveillance or signatures of humans, refers to technologies that measure and analyzes human physical and/or behavioral characteristics for authentication, identification, or screening purposes. Examples of physical characteristics include fingerprints, DNA, iris/retina, and facial. While the program was cancelled in 2003 due to wide criticism of potential abuses of domestic civil liberties, all elements of the program were continued under different names.

"Remote Brain Targeting/Remote Neural Monitoring is very much 'a reality'. However, obtaining proofs to support the testimonies of 'Targets' on the web is extremely difficult because there are International Intelligence Agency Agreements in place to suppress and deny any information relating to the use of this technology at all. It's categorized as 'Non-Lethal Weaponry' and is undoubtedly being used by sanctioning governments around the globe for purposes of surveillance, the suppression of those they deem to be any form of threat by means of torturing them, or, simply just for the purpose of scientific experimentation. It's highly illegal and offers / represents, via our own government institutions, a far greater threat to our civilian populations than any terrorist threat. There's also a media blackout in place concerning the covert use of the technology too. These situations mirror themselves, and amazingly, this is true all around the world.

The powerful capability to cause grave illness or worse by sugar coated descriptive terms designed as a play on words, such as "non-lethal" cannot be overlooked as intentional. The use of the technology has serious health issues and life-threatening man-made illness associated with the synthetic technology. Not only is the mind destroyed but also the body. Had not the United States learned anything from the Soviet Moscow incident? The Electronic Harassment, Directed Energy Weapon and Radio Frequency attack symptoms are:

FIRST STAGE:

feeling a shock as one drifts off to sleep, or seeing, seeing lights as one drifts off to sleep, unaccountable increased heart rate, nervousness, irritability, aggravation reactions, erection dysfunction, compressing the vertebrae noted upon waking, sensitivity to sound, depression, minor spasms and Charlie horses, inability to concentrate, loss of memory, night sweats, sensitive ear to touch, upon waking finding that the muscles of the back are tingling as if electrified, ringing in one or both ears and or hear tone bursts, GERD (Difficulty in Swallowing), Arthritis, eye sight becomes blurred, teeth snap together when drowsy, loss of sleep, nails become wavy, water weight and cellulite accumulations on the upper thigh, and buttocks from cellular fluids displaced generally from the head and upper body especially in women, where those fluids find ready redistribution in those areas, changes in the tone of the internal voice, waking up in extreme pain, and arthritis condition, spontaneous tearing without emotional thinking, mild pressure in the head, a cloudy feeling, increased need to urinate during the night, feeling puffs of air on the face and back, such may be a demodulation of energy on the skins surface, sensation that blood is trickling into localized areas of the brain and other parts of the body, this feeling is akin to the impression that these areas were devoid of blood, unaccountable increased heart rate just before drifting off to sleep causing the person to wake up

SECOND STAGE:

burning sensation on the skin, loss of hair, feeling of temporary heating of the head (demodulating RF effect), rashes, narcoleptic reactions, sleepiness, sleep only after exhaustion, virtual insomnia, miscarriage, gaunt face, facial wrinkles, losing skin turgor, hair breaks from rapid microwave heat on hair causing increase fracturing of hair shafts from expansion of water via humidity, losing control (unexplained anger), heating of body (body feeling as if it is being cooked), waking after sleep feeling as though you did not sleep, children's symptoms can manifest themselves as ADD and Hyperactivity

THIRD STAGE:

effects of exposure, loss of coordination, accidents from sleeplessness, damage to eyesight, atrophy of the muscles, heart valve damage, loss of weight and/or weight gain, nausea, sensitivity to sound, decreased dexterity, seizure, choking, vivid dreams, lost time (Alzheimer type symptoms), change of mood, heart attack, the personality becomes quiet, a lack of thought takes place, decrease in mental activity, limbs jerk typically as one is trying to sleep, weakness, lethargy or hyperactivity, forgetfulness, constipation

FOURTH STAGE:

Syndromes, disease, insanity, heart attack, stroke and death.

CHAPTER ELEVEN

"Does Big Brother exist?" "Of course, he exists. The Party exists. Big Brother is the embodiment of the Party." "Does he exist in the same way as I exist?" "You do not exist."

- George Orwell (1984)

CONCLUSION

In 1945, George Orwell wrote a futuristic novel called 1984. It was written during when mind control technology testing, research and development by definitive objective were occurring all over the world in secret laboratories escalated by the Arms Race between the United States and the Soviet Union during the Cold War. Today, it is plain to see that the foresight of George Orwell's "1984," a fictional account of a totalitarian mentality of government as chilling and prophetic. It is even more so prophetic, as a now fully implemented global technological capability continues to unfold resulting from varied and many, testing programs throughout the history of the United States and the other nations.

Published Mind Control - Open Literature continues to substantiate that various forms of electromagnetic weaponry today had already been in the hands of the military and highly placed intelligence agencies and technological operation/fusion centers. The High Frequency Active Auroral Research program or HAARP Atmospheric

Research Facility in Alaska is fully active. GWEN Towers are functional as is the digital delivery system called Silent Subliminal Spread Spectrum or (SSSS). All are subliminal message carrying technologies. However, most officials will not admit to its capability or use, or even that it can be used in this way? Don't count on correct information if asking. Local law such as the New York and Los Angeles Police Departments now have a fully operational state-of the-art-operation centers which have similar capabilities to Federal fusion centers, for example the LAPD's Real Time Analysis Command Division.

Across the nations, police are equipped with microwave weapons ranging from acoustic "sonic scream" devices which and different versions of the "active denial systems," such as even the handheld, ray and stun gun. The Army Telepathic Ray Gun or MEDUSA can be used from adjacent apartments to transmit painful rays and carried the sender'. It can also be used and as a two-way voice transmitter. Today electromagnetic technology is in full use and used as a matter of procedure. See through the wall technology such as the Xaver 800 developed by the military is available at the Federal, state and local levels also. X-ray vision allows through the walls viewing from neighboring apartments, and can also be use for officer safety, coexisting with real time situational awareness monitoring overhead from an operation/fusion center via satellite imagery. 13,000 objects orbit our planet today with 7% operation for military operations, etc. Legally the use of these technologies has been neatly packaged by direct intention for use today by sanctioning laws, regulations, also such as DOD Regulation 5240.1.R., as enforcement tools under the guise of riot or crowd control testing which has ultimately legitimize so-called, non-lethal weapons.

There are many, many people all over the world today cognizant of what is happening in covert programs, such as the United Nations, the United States Government, Congress, Senate, and the Intelligence Community, Federal, state and local agencies and also many governments make contribution through various efforts. Canada, the United Kingdom, Japan and Sweden are but a few locations where citizens are reporting these types of covert, heinous victimizations and abuses and crying out to deaf ears. There are major universities assisting in continued testing and research programs and developments. Some public citizens are aware of what is happening today, due mostly to personal experience, or by involvement with a person or persons connected to them and experiencing this phenomenon.

A diagnosis of Schizophrenia, delusional or psychosis is common for everyone, and I do mean everyone, without exception speaking up.

The DSM-IV-TR Diagnostic Criteria details symptoms typically occurring during a specific onset period in a person's life for schizophrenia. The documented onset years for paranoid schizophrenia, etc., for a woman, for example, is between the ages of 25 to age 30 and usually not after in most cases:

A Delusion Disorder is described below:

This disorder is characterized by the presence of non-bizarre delusions which have persisted for at least one month. Non-bizarre delusions typically are beliefs of something occurring in a person's life which is not out of the realm of possibility. For example, the person may believe their significant other is cheating on them, that someone close to them is about to die, a friend is really a government agent, etc. All of these situations could be true or possible, but the person suffering from this disorder knows them not to be (e.g., through fact-checking, third-person confirmation, etc.).

People who have this disorder generally don't experience a marked impairment in their daily functioning in a social, occupational or other important setting. Outward behavior is not noticeably bizarre or objectively characterized as out-of-the-ordinary.

The delusions cannot be better accounted for by another disorder, such as schizophrenia, which is also characterized by delusions (which are bizarre). The delusions also cannot be better accounted for by a mood disorder, if the mood disturbances have been relatively brief.

The common disbelief of the uninformed public is expertly nurtured by those operating in these highly secretive programs. Discrediting is a very important and valuable proponent which serves as an effective means to successfully keep these covert programs under wraps as long as possible and allowing them to thrive. A misdiagnosis enables the continued abuse and cover up of electronic harassment in Psychological Operations, regarding the Targeted Individual. Psychological Electronic technology is capable of manipulation, influencing, and pushing a person over the edge or into committing suicide or by energy weapon irradiation, necrosis, cancer and death. A person can be targeted for years. This fact alone lends credibility to the reality of human guinea pig testing. It appears that individuals having a history of substance abuse can be typical targets for testing purposes due to the ability to redirect their complaints of Remote Brain Targeting as drug induced psychosis by design. Activist, Whistleblowers, dissidents, or just angering the wrong person, as seen with Harlan Girard on the Jesse Ventura Conspiracy Theory, Brain Invader episode, TruTV, can result in life long torment also.

Again, people with a Delusional Diagnosis are recognized a person(s) having delusions involving real-life situations that could be true, such as being followed, being conspired against, or having a disease, or even a law enforcement enacted covert investigation. In fact, the Martha Mitchell Effect, a term used by psychologist, defines this dynamic in which real life situations or circumstances that cannot be verified by a clinician has resulted in misdiagnosis of many, many credible average individuals and even highly educated individuals.

PSYCHOLOGISTS AND THE MITCHELL EFFECT

John Mitchell was the Attorney-General during the Nixon administration.

His wife - Martha Mitchell - told her psychologist that top White House officials were engaged in illegal activities. Her psychologist labeled these claims as caused by mental illness.

Ultimately, however, the relevant facts of the Watergate scandal vindicated her.

In fact, psychologists have now given a label - the "Martha Mitchell Effect" - to "the process by which a psychiatrist, psychologist, or other mental health clinician mistakes the patient's perception of real events as delusional and misdiagnoses accordingly".

The authors of a paper on this phenomenon (Bell, V., Halligan, P.W., Ellis, H.D. (2003) Beliefs About Delusions. The Psychologist, 6 (8), 418-422) conclude:

Sometimes, improbable reports are erroneously assumed to be symptoms of mental illness [due to a] failure or inability to verify whether the events have actually taken place, no matter how improbable intuitively they might appear to the busy clinician.

In other words, psychologists who haven't taken the time to examine for themselves the claims of their patients will tend to label as delusional anything which they "intuitively" feel is improbable.

Many psychologists - just as Martha Mitchell's - will tend to assume any claim of conspiracy is improbable. However, conspiracies are actually common occurrences which are well-recognized by the law.

Psychologists are even more apt to label government conspiracies as improbable. However, as Martha Mitchell's psychologist learned, they do happen. Watergate, for example, was a conspiracy.

Psychologists who have attempted to label as delusional those who raise the possibility of government conspiracies do not have even a basic understanding of the Martha Mitchell Effect, or have not examined whether or not there is any factual basis for their patient's claims.

"Obviously, some people are delusional, and see conspiracies where none exist. But it is equally true that when millions of scientists, military leaders, historians, legal scholars, intelligence officials and other rational people say the government is lying, psychologists who dismiss similar claims by their patients are falling prey to the Martha Mitchell Effect. They are too busy and/or arrogant to actually examine their assumptions as to whether or not the claims which feel improbable to them are true."

The "Martha Mitchell Effect," is a psychiatrist mistakenly identifying a patient's extraordinary claims as delusions, despite their veracity, was later named after her.

Wikipedia reports that the Mitchell's separated in 1973. The marriage, apparently, unable to sustain the strain of the resulting eruption of Mrs. Mitchell's factual claims of corruption in the Nixon White House, nor hold up to the historical aftermath of the destruction and demise of the Nixon Administration and Presidential Impeachment. Her contacting reporters when her husband's role in the scandal became known had earned her the dubious title, "the Mouth of the South".

Wikipedia also states that Richard Nixon was later to tell interviewer David Frost (in September 1977 on Frost on America) "If it hadn't been for Martha Mitchell, there'd have been no Watergate."

Mrs. Mitchell had also insisted she had been held against her will in a California hotel room and sedated to prevent her from making the controversial phone calls to the news media. Because of her allegations, she was discredited and even abandoned by most of her family, except her son Jay. Nixon aides even leaked to the press that she had a "drinking problem In 1976, in advanced stages of myeloma, Mrs. Mitchell slipped into a coma and died in New York at age ripe young age of 57. She is buried in the Bellwood Cemetery in Pine Bluff.

The active role of health care professionals cannot be denied, whether knowingly or unknowingly in efforts to deny the existence of mind control. This is specifically relevant in the field psychiatry and psychology. By all account, there appears to be a joint conspiracy to discredit people speaking up regarding, factual personal experiences of these various technologies. The diagnosis of those speaking up across the board coincides with a quick diagnosis of delusions or psychosis disorders without investigations or in fairness to some, the ability to investigate if they could if what the patient is saying is factual.

In 2008, Dr. Ralph Hoffman, a psychiatric professor at Yale who studies delusions documented his belief that numerous individuals experiencing surveillance and psychotronic technology, globally, was related to little more than the results of numerous websites of numerous victims, activist, whistleblowers, including those used in factual testing programs or as guinea pigs speaking up. What continually amazed me is the inability to accept even the remotest possibility by some, that the technology exists, thus denying that actual laws, and factual patents, and historic inventors. What even more amazing is the fact that many of these programs went forth hand-in-hand with the Association of Psychiatry, and many Neuroscientist.

While delusional thinking does exist, the common assessment that if a person believes themself targeted in this manner, that they must be mentally ill is highly questionable in light of the capabilities of these brilliant advancements now highly perfected. Studies have proven definite microwave bio-effect congruence with Schizophrenia. These types of targeting have an identifiable pattern and a structure to the harassment that has been obviously refined and highly turned by design and years of unawareness to their usage. From my person experience, diagnosis is intended to make the target look crazy, which thereby serves to isolate the individual making the claim, which makes the person even more vulnerable and prey to ruthless predators and manipulators lacking the consciousness to stop even in the face of failure to obtain their objective. Herein lays my extreme problem with covert targeting, among other reasons. Sadly, the activity is so traumatizing that many otherwise "mentally healthy" individuals can easily develop mental health issues as a result of the stalking by satellite surveillance. The individuals manning the technology vindictively seem to go to great lengths seeking their results and are willing to do or say anything to achieve them. These types of surveillances are the result of a spiritual disease and moral decay which is unsurpassed in the history of humanity. Targeting's are result oriented situations whether for testing purposes to provide data, for law enforcement, revenge, manipulation, or to silence a whistleblower or even a political opponent. As a result, mental illness is not an indicator of whether or not the activity is actually taking place.

"Take away the words "mind control," which most feel distasteful or bizarre, and think of someone being manipulated by suggestions or told that what they know or what they are experiencing is not the truth, that it is their imagination. It is understandable, that the use of the term "mind control" would make the public think of people in a stupor. However, the fact remains that some, myself included can recognize what is happening to them and are able to spread the word though at a cost. True also, yes, there are people in today's society that believe in aliens and being abducted by aliens, but they do not belong to the group of people that are passionately trying to alert the public to the abuses escalating on citizens via these covert, subliminal, manipulative activities and weapons."

Igor Viktorovich Smirnov was a controversial Russian scientist best known for his role in Soviet-era mind control research as well as an obscure field of human behavior study he called "psycho-ecology". Smirnov has been characterized in the media as a Rasputin-like character with "almost mystical powers of persuasion." According to his wife Rusalkina, the Soviet military enlisted Smirnov's psycho-technology in the 1980s to combat the Mujahideen and treat post-traumatic stress syndrome in Russian soldiers during the war in Afghanistan.

After the dissolution of the Soviet Union, Smirnov transitioned from military research to the treatment of patients with drug addiction and mental problems. He founded the Psycho-technology Research Institute at the Peoples' Friendship University of Russia to work on ideas like "psycho-correction," a term he used to denote the use of subliminal messages to alter a subject's will, or even modify a person's personality without their knowledge. Struggling to gain acceptance in the West, the Institute caught the attention of former science-fiction writers turned Pentagon consultants Chris Morris and Janet Morris, who in 1991 tried unsuccessfully to market the technology to the United States military. Smirnov gained brief fame in the U.S. in 1993 when the FBI consulted with him in hope of ending the Waco Siege. Smirnov proposed blasting scrambled sound over loudspeakers to persuade David Koresh to surrender, however the FBI declined the plan. Smirnov died in November 5th 2004, leaving his wife to run the Institute.

In May 2009, the U.S. Department of Homeland Security announced plans to award a contract for testing of an airport screening system based partly on Smirnov's concepts to ManTech SRS Technologies in association with Northam Psycho-technologies, a Canadian company acting as distributor for the Psycho-technology Research Institute.

Igor Smirnov, in Moscow, said in regard to this technology: "It is easily conceivable that some Russian 'Satan', or let's say Iranian (or any other 'Satan'), as long as he owns the appropriate means and finances, can inject himself (intrude) into every conceivable computer network, into every conceivable radio or television broadcast, with relative technological ease, even without disconnecting cables. You can intercept the (radio) waves in the ether and then (subliminally) modulate every conceivable suggestion into it.

If this transpires over a long enough time period, it accumulates in the heads of people. Eventually they can be artificially manipulated with other additional measurements, to do that which the perpetrator wants them to do. This is why such technology is rightfully feared."

Today, the numbers continue to rise among people globally voicing their outrage of these types of covert, sanctioned, operations. Even if a person is targeted by law enforcement, the individual's situation bypasses the justice system as a place designed for addressing allegations and the target become the target of malicious prosecution and abuse of process relentlessly. The military programs are mainly technology testing programs and a person can be put into one of these programs by simply angering the wrong person.

When a person is targeted they are swimming upstream, in turbulent waters, against a powerful tide which can leave them standing alone. The technology can and is being used to manipulate those around them creating chaos, while some individual or individuals sit behind supercomputers. Some have Targeted Individuals have given in, gave up the difficulty fight for credibility and have simply learned to cope with the incessant technological telepathy around the clock, the electromagnetic energy weapon assaults as torture sadly. They had no choice but to try and live with it and cope or be committed to a mental hospital.

The people working these programs are employees. They do this for a living and work in shifts around the clock using technology which desensitizes them to the pain and suffering giving the feel of playing an advanced computer game. If I can stress anything as vital here, I must stress that the psychosis diagnosis of Targeted Individuals is the ace-in-the hole for continued testing of technology on human subjects around the world legally.

George Orwell, 1984

Now I will tell you the answer to my question. It is this. The Party seeks power entirely for its own sake. We are not interested in the good of others; we are interested solely in power, pure power. What pure power means you will understand presently. We are different from the oligarchies of the past in that we know what we are doing. All the others, even those who resembled ourselves, were cowards and hypocrites. The German Nazis and the Russian Communists came very close to us in their methods, but they never had the courage to recognize their own motives. They pretended, perhaps they even believed, that they had seized power unwillingly and for a limited time, and that just around the corner there lay a paradise where human beings would be free and equal. We are not like that. We know what no one ever seizes power with the intention of relinquishing it. Power is not a means; it is an end. One does not establish a dictatorship in order to safeguard a revolution; one makes the revolution in order to establish the dictatorship. The object of persecution is persecution. The object of torture is torture. The object of power is power. Now you begin to understand me.

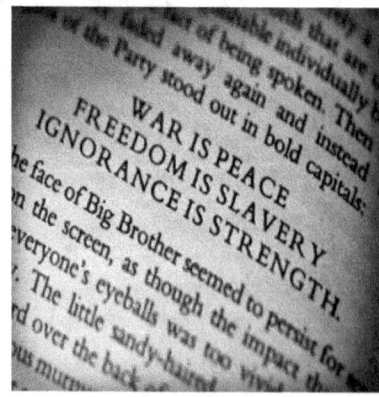

George Orwell, 1984

In the end, the Party would announce that two and two made five, and you would have to believe it. It was inevitable that they should make that claim sooner or later: the logic of their position demanded it. Not merely the validity of experience, but the very existence of external reality, was tacitly denied by their philosophy. The heresy of heresies was common sense. And what was terrifying was not that they would kill you for thinking otherwise, but that they might be right. For, after all, how do we know that two and two make four? Or that the force of gravity works? Or that the past is unchangeable? If both the past and the external world exist only in the mind, and if the mind itself is controllable—what then?

The End or The Beginning of Awareness?

If you enjoyed reading Book I, Remote Brain Targeting by Renee Pittman you will certainly enjoy reading the other five books in the "Mind Control Technology" book series which are available at the links below, in print, on amazon.com and Amazon Kindle as EBooks:

- Book I - "Remote Brain Targeting"

- Book II - "You Are Not My Big Brother - Menticide"

- Book III - "Covert Technological Murder" - Pain Ray Beam."

- Diary of an Angry Targeted Individual – Mind Invasive Technology"

- Book V - "The Targeting of Myron May"

- Book VI = "Deceived Beyond Belief – The Awakening"

It appears that "Big Brother" has created, clean shaven, uniformed, suited, covert technological monsters.

Website: http://bigbrotherwatchingus.com/
You may email the author at:
big.brotherwatching@live.com

IMAGE SECTION

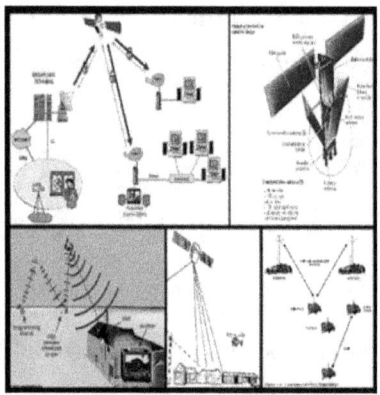

Satellites can not only see through walls but also deep into the earth's surface. Satellite imagery is sent back to relevant computers in state-of-the-art military and law enforcement operation/fusion centers and is cost free to a requesting agency. It can also be sent to portable equipment such as the Panasonic "Tough Book" laptop weapon and real-time system.

Remote Brain Targeting

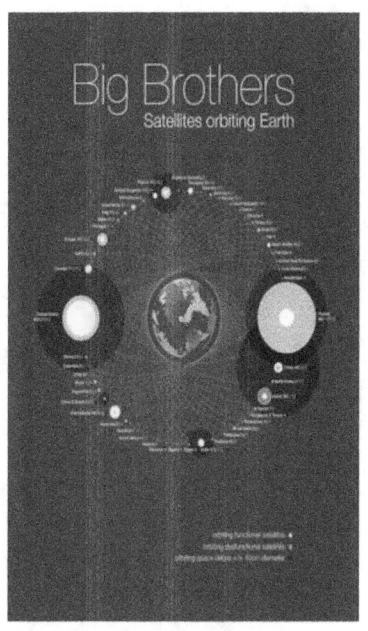

Numerous satellites are orbiting earth and the numbers are growing rapidly.

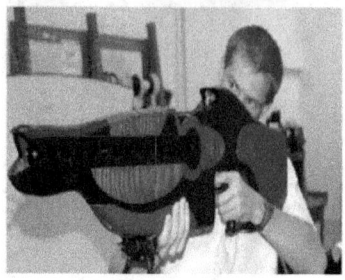

The, MEDUSA, above and below, is a military telepathic ray gun and directed energy weapon. It available to all levels of law enforcement also.

Remote Brain Targeting

THE FACTS

- These technologies are not freely available to honest law enforcers, only to spies and other criminals; their "associates".
- These "associates" include military madmen, political puppets, News Nazis and other organised crime figures (e.g. almost the entire entertainment industry is a protected organised crime operation).
- The apparent respectability of many public figures is only an illusion, like the false impression that world events are random. Images and scenarios are manufactured to deceive the public and serve the corrupt masters of the agencies.
- The agencies stock in trade includes lies, rumours and terror. Spook (ghost) writers are planted and recruited in politics, the media etc. Most public figures will say, do, read, sing etc whatever they're told/paid to. They don't care about the source (often surveillance) or targets (innocent victims).

THE TARGETS

- Anyone challenging powerful criminals ideologically, politically etc is targeted.
- These criminals run lucrative, protected operations, the drug trade, arms trade, systematic theft of intellectual property, human experimentation, terrorism etc.
- You cannot criticize warmongers, drug barons or other oppressors without repercussions.

THE SECRECY

- "National Security" and official secrets legislation protect the perpetrators.
- Media complicity keeps the information out of the public domain e.g. one IBM Blue gene computer can process more information than six times the worlds total population (227 trillion calculations per second).
- Others who knew all of this are silenced; through satellite oppression or worse.
- The criminals responsible delude themselves by equating their best interests with those of society in general. They thereby justify (to themselves) the mistreatment of free thinkers who question their crimes using the technologies listed over the page. All of those involved belong in prison.

FOR FURTHER INFORMATION SEE www.surveillanceissues.com

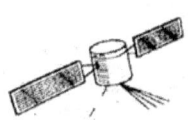

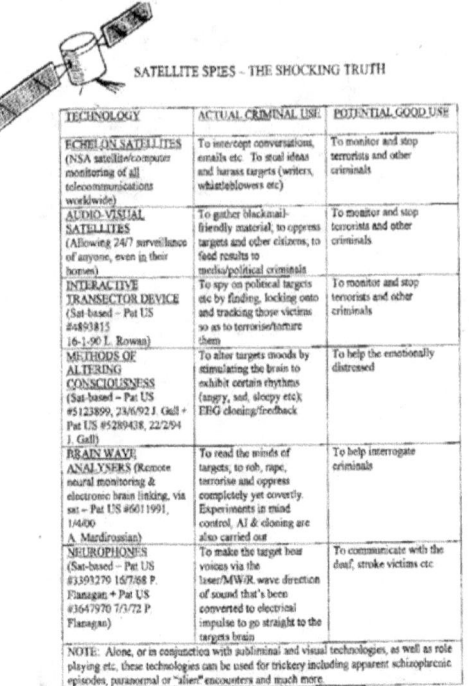

Directed Energy Weapons can be deployed in many ways at a Targeted Individual to include aircraft and satellite energized drones.

Remote Brain Targeting

NOTE: Electronic Harassment technology today is also said to be in the hands of covert harassment groups and government contractors. For example, Lockheed Martin was awarded the High-Power Microwave Energy Weapon Contract. Raytheon, Militech and Olin Corporations make portable compact microwave weaponry for example..

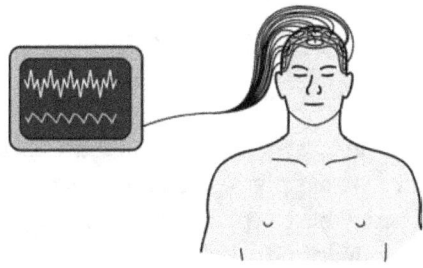

Electroencephalogram mind reading is done by patented technology without electrolodes

How is it done?

The magnetic field around the head, the brain waves of an individual can be monitored by satellite.

The transmitter is therefore the brain itself just as body heat is used for "Iris" satellite tracking (infrared) or mobile phones, RFID chips or bugs can be tracked as "transmitters." In the case of the brain wave monitoring the results are then fed back to the relevant computers.

Monitors then use the information to conduct "conversation" where audible Neurophone input is "applied" to the target / victim.

United States Patent 5527352 Time Focused Induction of Preferential Necrosis. This patent creates a slow/kill effect from non-ionizing radiation. Microwave burn leads to necrosis eventually when the radar laser beam is focused on body parts 24/7 intentionally by Microwave Directed Energy Weapons. I have lost two hips before my knees became a target.

When targeted, these groups take great pleasure in insuring crippling before gym activities or simply just running errands or as soon as a person gets out of bed. These repeated attacks lead to microwave burn necrosis and are no longer "non-lethal" as deceptively termed.

Danger Room "Science"

Army Funds 'Synthetic Telepathy' Research

By Noah Shachtman, 08.18.08, 12:28 PM

"The Army has given a team of University of California researchers a $4 million grant to study the foundations of "synthetic telepathy." But unlike old-school mind-melds, this seemingly psychic communication would be computer-mediated. The University of California, Irvine explains…"

Thousands Targeted in USA and millions globally are reported worldwide to be suffering electronic harassment directed energy weapon attacks and covert harassment from psychotronic technology.

The National Security Agency mind/brain interface/supercomputer operation center, in Utah will be completed by fall 2013.

LOCATIONS OF GWEN TOWERS IN THE US

When doubts arose regarding the threat of electromagnetic pulse to permanently shut down communications, only 58 of the originally planned 240 **GWEN towers** were built. In 1994 a defense appropriations bill banned new **towers** from being built, and shortly after, the **GWEN** program was cancelled by the Air Force..

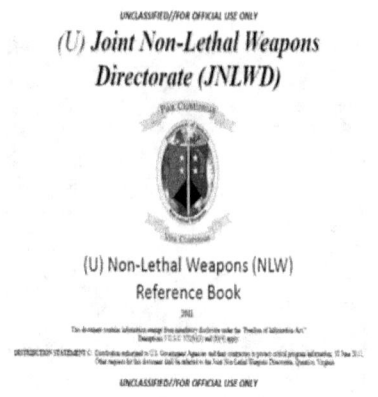

The term "non-lethal" is very deceptive and does not express the full range and scope or capabilities of these technologies.

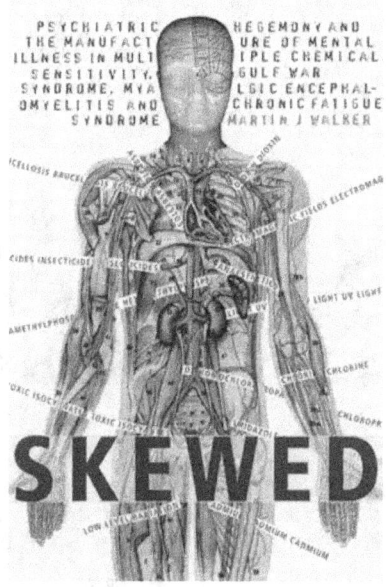

Manufactured mental illness continues to be an important strategic effort in use to silence

Targeted Individuals and for discrediting Targeted Individuals when speaking out and is fully effective for now. Knowledge of technological advancements is vital today.

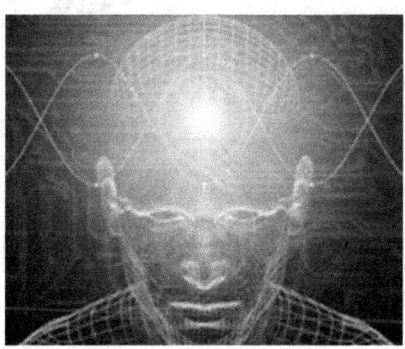

Synthetic Telepathy, Voice to Skull, Neural Decoding Remote Neural Monitoring, technology and a series of mind control patents are here to stay and legalized for military and law enforcement non-consensual testing.

people all over the world are saying the same thing. It's time to listen!

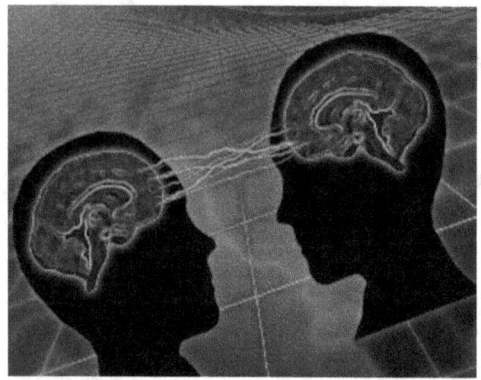

Voice to Skull, neural decoding, synthetic or artificial telepathy, or Frey Effect, etc., is the capability to beam voices directly into the human brain and it can be done from a state-of-the-art operation center.

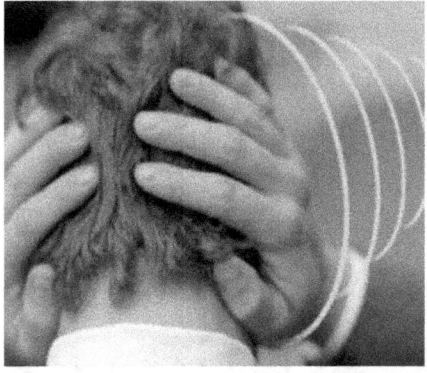

There is nothing non-lethal about these psychophysical systems and devices when used 24/7 to destroy lives covertly

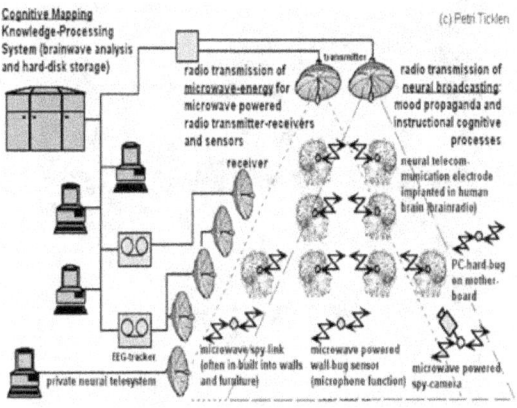

Brainwave analysis allows 24/7 Synthetic Telepathy harassment

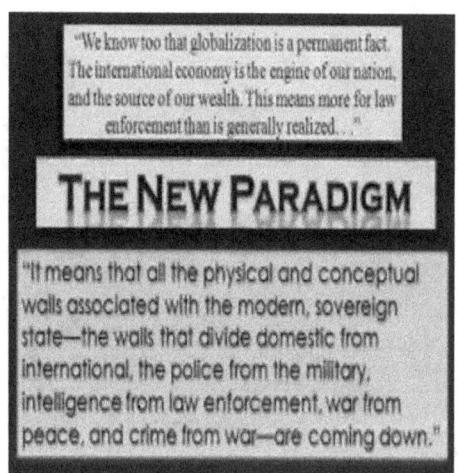

Law enforcement AND the military Joint Targeting Operations

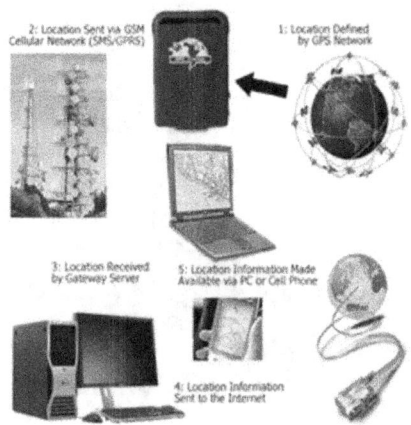

Point to point communication

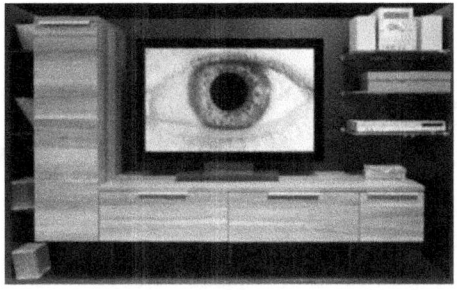

Computers, cellular phone, bank cards, and even televisions today have miniscule Radio Frequency Identification RFID chips also allow remote viewing and tracking.

Biometric signature reading technology, again portable, handheld, land, sea and space-based, tracks by iris, gait, DNA, etc.

Violations of Fourth Amendment Rights continue to be an immoral unaddressed issue by operators of this technology which allows in home real time viewing and covert energy weapon attacks inside a once private dwelling from nationwide advanced operation centers and portable technology as close as next door.

REFERENCES

- (gleaned from excerpts in Chapter Four – Courtesy of John J. McMurtrey)

- Frey AH. "Auditory System Response to Radio Frequency Energy" Aerosp Med 32: 1140-2, 1961.

- Elder JA and Chou CK. "Auditory Responses to Pulsed Radiofrequency Energy" Bio electromagnetics Suppl. 8: S162-73, 2003.

- Lin JC. "Auditory Perception of Pulsed Microwave Radiation" In: Gandhi OP (ed.) Biological Effects and Medical Applications of Electromagnetic Energy Prentice Hall, Englewood Cliffs, NJ, Chapter 12, p 278-318, 1990.

- Justesen DR. "Microwaves and Behavior" American Psychologist 392(Mar): 391-401, 1975. Accessed 3/8/05 at Microwaves and Behavior

- Oskar KJ. "Effects of low power microwaves on the local cerebral blood flow of conscious rats" Army Mobility Equipment Command Report # AD-A090426, 1980. Abstract accessible 4/8/05. Available from NASA Technical Reports.

- Kohn B. "Communicating Via the Microwave Auditory Effect" Defense Department Awarded SBIR Contract # F41624-95-C9007, 1993. Contract abstract at:

- Brunkan WB. Patent # 4877027 "Hearing system" USPTO granted 10/31/89.

- Leyser R. Patent # DE10222439 "Microwave hearing device uses modulated microwave pulses for providing induced sound warning directly within head of deaf person" Federal Republic of Germany Patent and Trademark Office published 12/11/03. Abstract accessed 12/14/03

- Original German Document accessed 12/14/04

- O'Loughlin JP and Loree DL. Patent # 6470214 "Method and device for implementing the radio frequency hearing effect" USPTO granted 10/22/02.

- O'Loughlin JP and Loree DL. Patent # 6587729 "Apparatus for audibly communicating speech using the radio frequency hearing effect" USPTO granted 7/1/03.

- "Surveillance Technology, 1976: policy and implications, an analysis and compendium of materials: a staff report of the Subcommittee on Constitutional Rights of the Committee of the Judiciary. United States Senate, Ninety-Fourth Congress, second session, p 1280, 1976. US GOV DOC Y 4.J 882:SU 7/6/976.

- Castelli CJ. "Questions Linger about Health Effects of DOD's 'Non-Lethal Ray'" Inside the Navy 14(12): 1-6, 2001. Full text 4/7/05 accessible at

- Center for Army Lessons Learned Thesaurus

- Apparently periodically terms are added to this Thesaurus and the URL for this entry may change. If the link is broken go to the thesaurus at:

- Voice to Skull. Since the present article has been posted on the Internet, the entry has been programmed so that it cannot be printed. The Federation of American Scientists Project on Government Secrecy has made note of this: "Voice to Skull: More

Army Web Shenanigans" Secrecy News, vol. 2004, issue 64, July 12, 2004, the last item

- Krawczyk G. "CIA Using Old Tricks Again" Nexus Magazine, Oct/Nov, 2(22): 9, 1994.

- McMurtrey J. "Inner Voice, Target Tracking, and Behavioral Influence Technologies" in press 2005.

- Flaum M and Schultz SK. "The Core Symptoms of Schizophrenia" Ann Med 28(6): 525-31, 1996.

- Nayani TH and David AS. "The auditory hallucination: a phenomenological survey" Psychol Med 26: 177-89, 1996.

- Hubl D, Koenig T, Strik W, Federspiel A, Kreis R, Boesch C, Maier SE, Schroth G, Lovbald K, and Dierks T. "Pathways that Make Voices: White Matter Changes in Auditory Hallucinations" Arch Gen Psychiatry 61: 658-68, 2004.

- Isselbacher KJ, Adams RD, Brunwald E, Petersdorf RG, and Wilson JD (eds.) Harrison's Principles of Internal Medicine Ninth Ed., McGraw-Hill, New York, p 150, 1980.

- American Psychiatric Association DSM-IV Task Force. Diagnostic and Statistical Manual of Mental Disorders Fourth Edition (DSM-IV-TRTM) American Psychiatric Association, p 297-343, 2000.

- Gandhi OP. "Electromagnetic Fields: Human Safety Issues" Ann Rev Biomed Eng. 4: 211-34, 2002.

- Steneck NH. The Microwave Debate. MIT Press, Cambridge, London, 1984; p 93, 181-9, 208

- United States Congress, Senate Committee on Commerce, Science and Transportation. Microwave irradiation of the U.S. Embassy in Moscow: review of its history and studies to determine whether or not related health defects were experienced by employees assigned in the period 1953-1977. US Government Printing Office, 1979

- Smith CW, Best S. Electromagnetic Man. J.M. Dent & Sons Ltd., London, 1989; p 211, 233, & 235

- Johnson Liakouris AG. Radiofrequency (RF) Sickness in the Lillenfeld Study: An Effect of Modulated Microwaves? Arch Environ Health 1998;53(3):236-38

- Goldsmith JR. Epidemiological Evidence of Radiofrequency Radiation (Microwave) Effects on Health in Military, Broadcasting, and Occupational Studies. Int J Occup Environ Health 1995; 1:47-57

- Goldsmith JR. Where the trail leads . . . Ethical problems arising when the trail of professional work leads to evidence of cover-up of serious risk and mis-representation of scientific judgment concerning human exposures to radar. Eubios Journal of Asian and International Bioethics 1995; 5:92-4

- Schiefelbein S. The Invisible Threat: The Stifled Story of Electric Waves. Saturday Review, 1979 Sept 15: 16-20, p 17

- Brodeur, P. The Zapping of America. Norton, New York, 1977; p 105

- Cesaro RS. Memorandum for the Director, Advanced Research Projects Agency, Subject: Justification Memorandum for Pandora, 15 Oct 1965.

- Cesaro RS. Memorandum for the Director, ARPA Subject: Project Pandora -- Initial Test Results Advanced Research Project Agency (letterhead) with marked out "Limited Access" and "Eyes Only" in addition to "Top Secret", 15 Dec 1966 (stamp).

- CALIBRATED MICROWAVE FACILITY AT WALTER REED ANECHOIC CHAMBER, Irradiated Test Section 4' X 2' X 1 1/2'. Accessed 8 Nov 2008.

- Cesaro RS. Memorandum for the Record, Subject: Project Pandora - Initial Test Results, Reference: PANDORA-BIZARRE Test

Results - Memo dated 15 Dec 66, 20 Dec 1966 (stamp), Advanced Research Projects Agency (letterhead) also marked "Limited Access" and "Eyes Only" as well as "Top Secret."

- Cesaro RS. Memorandum for Director, Defense Research & Engineering, Subject: Project Bizarre, Advanced Research Project Agency (letterhead) with marked out "Top Secret" and "Limited Access," 27 Sept 1967 (stamp).

- Byron EV. Project Pandora (U), Final Report, Johns Hopkins Applied Physics Laboratory.

- Byron EV. Operational Procedure for Project Pandora Test Facility, Johns Hopkins Applied Physics Laboratory, Oct 1966.

- Peterson L, Kubis J, Baramack J, Hughes F, Pollack H. Memorandum To: Mr. R. S. Cesaro, ARPA, From: IDA Review Panel, Subject: Flash Report of Pandora/Bizarre Briefing (S), January 14, 1969.

- (FOIA release only detailed description of simulated signal matching less detailed reference with experimental results, and apparently prepared by Dr. Sharp for attachment to ref. **Error! Bookmark not defined.**). Appendix I, CALIBRATED MICROWAVE FACILITY AT WALTER REED ANECHOIC CHAMBER, Irradiated Test Section 4' X 2' X 1 1/2', as well as the following hand graphs.

- Becker RO. Cross Currents, Jeremy P. Tarcher, Inc., Los Angeles, 1990; 297-304

- Hyland GJ. Physics and Biology of Mobile Telephony. Lancet 2000; 356:1833-6

- Chanda M (Secretary to Dr. Pollack). Pandora Meeting of January 12, 1970 (U) Minutes Submitted by Lysle Petersen, Chairman. Mar 23, 1970.

- Kubis JF. On Evaluation of Data Associated with Pandora (Preliminary Report). 12/4/69.

- Chanda M. (secretary to Dr. H. Pollack). Letter to OSD/ARPA/Advanced Sensors, Mr. Cesaro. IDA Letterhead, with Jan. 12, Minutes appended.

- Rechtin E, Betta AW. Agreement Transfer of Project Pandora. Effective 1 July 1970.

- Cesaro RS, Meroney WH. Subject: Pandora - Preliminary Agreement for Transfer Plans to U.S. Army. 20 Mar 1970.

- Thomas JR, Finch ED, Fulk DW, Burch LS. Effects of Low-Level Microwave Radiation on Behavioral Baselines. Ann N Y Acad Sci 1975; 247:425-32

- Thomas JR, Schrot J, Banvard RA. Comparative Effects of Pulsed and Continuous-Wave 2.8-GHz Microwaves on Temporally Defined Behavior. Bio electromagnetics 1982; 3:227-35

- Raslear TG, Akyel Y, Bates F, Bell M, Lu S-T. Temporal Bisection in Rats: The Effects of High-Peak-Power Pulsed Microwave Irradiation. Bio electromagnetics 1993; 14:459-78

- D'Andrea JA, DeWitt JR, Emmerson RY, Bailey C, Stensaas S, Gandhi OP. Intermittent Exposure of Rats to 2450 MHz Microwaves at 2.5 mW/cm2: Behavioral and Physiological Effects. Bio electromagnetics 1986a; 7:315-28

- Maier R, Greter S-E, Maier N. Effects of pulsed electromagnetic fields on cognitive processes - a pilot study on pulsed field interference with cognitive regeneration. Acta Neurol Scand 2004; 110:46-52

- Becker RO, Selden G. The body electric: Electromagnetism and the foundation of life. New York: Quill William Morrow; 1985, p 319-320.

- Ritsher JB, Lucksted A, Otilingam PG, Gonzales M. Hearing voices: explanations and implications. Psychiatr Rehabil J 2004;27(3):219-227.

- Thomas P, Mathur P, Gottesman II, Nagpal R, Nimgaonkar VL, Deshpande SN. Correlates of hallucination in schizophrenia: A cross-cultural evaluation. Schiz Res 2007; 92:41-9.

- Kluft RP. First-rank symptoms as a diagnostic clue in multiple personality disorder. Am J Psychiatry 1987;144(3):293-7.

- Dunayevich E, Keck PE. Prevalence and description of psychotic features in bipolar mania. Curr Psychiatry Rep 2000;2(4):286-290.

- Adams RD, Victor M. Derangements of intellect, mood, and behavior, including schizophrenia and manic-depressive states. In: Isselbacher KJ, Adams RD, Brunwald E, Petersdorf RG, Wilson JD, editors. Harrison's principles of internal medicine, 9th ed. New York: McGraw-Hill; 1980, p 150.

- Franklin RD, Allison DB, Gorman RS (editors). Design and analysis of single-case research. Mahwah, New Jersey: Lawrence Erlbaum, Associates, Publishers; 1996.

- Weiguo D, Qunli W. Audio sound reproduction based on nonlinear interaction of acoustic waves. Journal of the Audio Engineering Society 1999;47(7/8):602-6.

- Bellin JLS, Beyer RT. Experimental investigation of an end-fire array. J Acoust Soc Am 1962;34(8):1051-4.

- Shealy WP, Eller AJ. Design and preliminary results of an acoustic parametric source in air. J Acoust Soc Am 1973; 54:297A.

- Bennett MB, Blackstock DT. Experimental verification of the parametric array in air. J Acoust Soc Am 1973; 54:297A.

- Widener MW, Muir TG. Experiments in parametric arrays in air. J Acoust Soc Am 1974;55(2):428-429A.

- Bennett MB, Blackstock DT. Parametric array in air. J Acoust Soc Am 1975;57(3):562-8.

- Yoneyama M, Fujimoto J-I, Kawamo Y, Sasabe S. The audio spotlight: An application of non-linear interaction of sound waves to a new type of loudspeaker design. J Acoust Soc Am 1983;73(5):1532-6.

- Aoki K, Kamakura T, Kumamoto Y. Parametric loudspeakers—Characteristics of acoustic field and suitable modulation of carrier ultrasound. Electronics and Communications in Japan 1991;74(Part 3, #9):76-81.

- Kamakura T, Aoki K, Kumamoto Y. Suitable modulation of the carrier ultrasound for a parametric loudspeaker. Acustica 1991; 73:215-17.

- Pompei FJ. The use of airborne ultrasonics for generating audible sound beams. Journal of the Audio Engineering Society 1999;47(9):726-31.

- Yang J, Sha K, Gan W-S, Tian J. Nonlinear wave propagation for a parametric loudspeaker. IEICE Transactions on Fundamentals 2004; E87-A (9):2395-2400.

- Kamakura T, Tani M, Kumamoto Y. Parametric sound radiation from a rectangular aperture source. Acustica 1994; 80:332-8.

- Satoh K. Sound reproduction devices and systems: Parametric speaker. In: Benson BK, editor. Audio engineering handbook. New York: McGraw-Hill; 1988, p 7.61-7.66.

- Moon B-C, Kim M-J, Ha K-L, Kim C-D. Radiation characteristics improvement of flexural type vibrator for parametric sound source in air. Japan Journal of Applied Physics 2002; 41:3458-9.

- Tan KS, Gan WS, Yang J, Er MH. Constant beam width beam former for difference frequency in parametric array. Proceedings of the ICASSP, IEEE International Conference on Acoustics, Speech, and Signal Processing 2003;5:361-4.

- Havelock DI, Brammer AJ. Directional loudspeakers using sound beams. J Audio Engineering Society 2000;48(10):908-16.

- Bush E. Meeting recap. October 29, 2002 – Alternative loudspeaker transducer technologies. Audio Engineering Society Los Angeles Section Meeting 2002 Oct 29; p 2. [Online] [cited 2008 Jan 3]

- Webb W. Directional beams refocus sound science. EDN 2003 May 15; p 30-4.

- Sparrow D. Best of what's new grand award winner: Hypersonic sound. Popular Science Dec 2002;61(6):94.

- Lowrey A. Apparatus and method of broadcasting audible sound using ultrasonic sound as a carrier. US patent # 6052336, 2000 Apr 18.

- Norris EG. Acoustic heterodyne device and method. US patent # 5889870, 1999 Mar 20.

- Croft JJ, Norris JO. Theory, history, and the advancement of parametric loudspeakers: A technology overview. American Technology Corporation, 2001-2003; Part # 98-10006-1100 Rev. E. [Online] [cited 2011 May 19]

- Norris EG. The creation of audible sound from ultrasonic energy – A fundamental paradigm shift. J Acoust Soc Am 1997; 101:3072.

- Davidson N, Lewer N. Research report No. 8. Bradford Non-Lethal Weapons Research Project (BNLWRP), Centre for Conflict Resolution, Department of Peace Studies, 2006 Mar; p 33-5. [Online] [cited 2011 May 19]

- American Technology Corporation. Military/Project Highlight. [Online] [Cited 2011 May 19]

- Davidson N, Lewer N. Research report No. 5. Bradford Non-Lethal Weapons Research Project (BNLWRP), Centre for Conflict Resolution, Department of Peace Studies, 2004 May; p 3 & 20. [Online] [cited 2011 May 19]

- Schollmeyer J. Pumping up the volume. Bulletin of the Atomic Scientists 2004 Nov/Dec;60(6):8-9.

- Government/Project Highlight. American Technology Corporation. [Online] [Cited 2011 May 19]

- Karp J. Hey, you! How about lunch? New laser like sound beams messages to shoppers, aid military in Iraq. Wall Street Journal 2004 April 1; Sect. B:1.

- American Technology Corporation's LRAD ™ instillations providing infrastructure protection to oil and gas industry. LRADs also being utilized for long range communications in Nigeria, India, and the Gulf Coast. Business Wire 2006 July 25.

- American Technology Corporation. LRAD-R Optional Sensor Inputs Brochure. Last modified 10/19/07. [Online] [Cited 2011 May 19]

- Holosonics announces Audio Spotlight® exhibit at Boston's Museum of Science. Holosonics Research Labs. Press release of 2003 Oct 6. [Online] [cited 2011 May 19]

- Audio Spotlight sound beam systems installed in General Motors display at Walt Disney's Epcot. Holosonics Research Labs. Press release of 2004 June 30. [Online] [cited 2011 May 19]

- Holosonics' Audio Spotlight technology installed at the Smithsonian. Holosonics Research Labs. Press release of 2003 Nov 6. [Online] [cited 2011 May 19]

- Lee JS. An Audio Spotlight creates a personal wall of sound. New York Times 2001 May 15; Sect. F:4.

- Sella M. The sound of things to come. New York Times Late Edition Final 2003 Mar 23; Sect. 6:34-9. [Online] [Cited 2011 May 19]

- Alexander JB. Future war: Non-lethal weapons in twenty-first-century warfare. New York: St. Martin's Press; 1999, p 101.

- Wild N. Hand-held ultrasonic through-the-wall monitoring of stationary and moving people. Government Technical Report # A857814, Nov 2003.

- Wild N, Doft F, Wondra J, Niederhaus S, Lam H. Ultrasonic through-the-wall surveillance system. Proceedings of SPIE 2002; 4708:106-13.

- Mind Justice (Formerly Citizens Against Human Rights Abuse), Director, Cheryl Welsh, 915 Zaragoza Street, Davis, CA 95616, USA. Website at http://www.mindjustice.org/

- Email is welsh@mindjustice.org

- Christians Against Mental Slavery, Secretary, John Allman, 98 High Street, Knaresbourough, N. Yorks HG5 0HN, United Kingdom.

- Website at http://www.slavery.org.uk/ Email is info@slavery.org.uk

- Moscow Committee for the Ecology of Dwellings, Chairman, Emile Sergeevne Chirkovoi, Korpus 1006, Kvrtira 363, Moscow Zelenograd, Russia 103575. .

- International Movement for the Ban of Manipulation of the Human Nervous System by Technologic Means, Founder, Mojmir Babacek, P. O. Box 52, 51101 Turnov, Czech Republic, Europe.

- Moscow Committee for the Ecology of Dwellings, Chairman, Emile Sergeevne Chirkovoi, Korpus 1006, Kvrtira 363, Moscow

- Zelenograd, Russia 103575. . Website at

- International movement for the ban of manipulation of the human nervous system by technologic means, founder, mojmir babacek, p. O. Box 52, 51101 turnov, czech republic, europe4

www.ingramcontent.com/pod-product-compliance
Lightning Source LLC
Chambersburg PA
CBHW071801080526
44589CB00012B/640